MODELING AND **CONTROL**
—— FOR ——
MICRO/NANO DEVICES AND SYSTEMS

EDITED BY

Ning Xi

Mingjun Zhang

Guangyong Li

CRC Press
Taylor & Francis Group
Boca Raton   London   New York

CRC Press is an imprint of the
Taylor & Francis Group, an **informa** business

AUTOMATION AND CONTROL ENGINEERING
A Series of Reference Books and Textbooks

Series Editors

FRANK L. LEWIS, Ph.D.,
Fellow IEEE, Fellow IFAC
Professor
The Univeristy of Texas Research Institute
The University of Texas at Arlington

SHUZHI SAM GE, Ph.D.,
Fellow IEEE
Professor
Interactive Digital Media Institute
The National University of Singapore

PUBLISHED TITLES

Modeling and Control for Micro/Nano Devices and Systems,
Ning Xi; Mingjun Zhang; Guangyong Li

Linear Control System Analysis and Design with MATLAB®, Sixth Edition,
Constantine H. Houpis; Stuart N. Sheldon

Real-Time Rendering: Computer Graphics with Control Engineering,
Gabriyel Wong; Jianliang Wang

Anti-Disturbance Control for Systems with Multiple Disturbances,
Lei Guo; Songyin Cao

Tensor Product Model Transformation in Polytopic Model-Based Control,
Péter Baranyi; Yeung Yam; Péter Várlaki

Fundamentals in Modeling and Control of Mobile Manipulators, *Zhijun Li;*
Shuzhi Sam Ge

Optimal and Robust Scheduling for Networked Control Systems, *Stefano Longo;*
Tingli Su; Guido Herrmann; Phil Barber

Advances in Missile Guidance, Control, and Estimation, *S.N. Balakrishna;*
Antonios Tsourdos; B.A. White

End to End Adaptive Congestion Control in TCP/IP Networks,
Christos N. Houmkozlis; George A Rovithakis

Robot Manipulator Control: Theory and Practice, *Frank L. Lewis;*
Darren M Dawson; Chaouki T. Abdallah

Quantitative Process Control Theory, *Weidong Zhang*

Classical Feedback Control: With MATLAB® and Simulink®, Second Edition,
Boris Lurie; Paul Enright

Intelligent Diagnosis and Prognosis of Industrial Networked Systems,
Chee Khiang Pang; Frank L. Lewis; Tong Heng Lee; Zhao Yang Dong

Synchronization and Control of Multiagent Systems, *Dong Sun*

Subspace Learning of Neural Networks, *Jian Cheng; Zhang Yi; Jiliu Zhou*

Reliable Control and Filtering of Linear Systems with Adaptive Mechanisms,
Guang-Hong Yang; Dan Ye

Reinforcement Learning and Dynamic Programming Using Function
Approximators, *Lucian Busoniu; Robert Babuska; Bart De Schutter; Damien Ernst*

Modeling and Control of Vibration in Mechanical Systems, *Chunling Du;*
Lihua Xie

CRC Press
Taylor & Francis Group
6000 Broken Sound Parkway NW, Suite 300
Boca Raton, FL 33487-2742

© 2014 by Taylor & Francis Group, LLC
CRC Press is an imprint of Taylor & Francis Group, an Informa business

First issued in paperback 2017

No claim to original U.S. Government works
Version Date: 20131014

ISBN 13: 978-1-138-07246-6 (pbk)
ISBN 13: 978-1-4665-5405-4 (hbk)

Visit the Taylor & Francis Web site at
http://www.taylorandfrancis.com

and the CRC Press Web site at
http://www.crcpress.com

Contents

Contents

Preface

It remains an especially daunting challenge for micro/nanoscale engineering to engineer ultra-fast and ultra-scale devices for implementation. Modeling and control play an essential role in advanced engineering and have contributed to many breakthroughs in modern technology revolutions. The need for modeling and control for micro/nanoscale devices and systems is fast emerging. We thus hope that our book on modeling and control will address the long-term challenge to advanced engineering for micro/nanoscale sensors, energy devices, and cellular and molecular systems. Unfortunately, few books in the literature offer an integrated view from theory to practice of this fast emerging subject. This book aims to provide an integrated view of this emerging field with a focus on theories and practices for practical implementation. The applications of the discussions are modeling and control over biosensors, energy devices, and molecular and cellular systems.

This book consists of nine chapters contributed by leading researchers in modeling and control for micro/nano devices and systems. Chapter 1 introduces the fundamental principles in modeling and design methods in quantum control theory as well as the major results obtained in physics, chemistry, and control sciences, and perspectives on future directions in this field. Chapters 3 and 4 focus on energy devices. Chapter 2 presents a modeling and simulation study of biosensors made of single nanowires for detecting biomolecules. Chapter 3 provides both continuous and statistical approaches for modeling and simulation of organic solar cells made of bulk heterojunctions in nanoscale. Chapter 4 discusses the optimization design of organic solar cells via multiscale simulation. Beginning with Chapter 5, the book shifts its focus to biological systems and bio-inspired systems. Chapter 5 discusses the viscoelastic property of the human epidermoid carcinoma cell line and investigates the cell modeling of the dynamic signaling pathway induced after epidermal growth factor simulation. Chapter 6 studies the cell tensegrity model by considering the cell body as an inhomogeneous structure, which consists of force-bearing elements, the cytoskeleton that is bounded by the cell membrane. Chapter 7 deals with modeling of swimming micro/nano-systems in a low Reynolds number with the goal of engineering micro/nanoscale propulsion systems for controlled drug delivery. Chapter 8 proposes integration of mathematical modeling and experimental techniques to connect local and systemic dynamics associated with cellular functions while affording alternative methods for uncovering the fundamental mechanisms behind the complex biological processes. Chapter 9 presents a hybrid control strategy for micro/nanoscale devices and systems that can be employed in broader applications.

We see the emergence of many theories and techniques in this rapidly growing field. It is difficult for any single book to capture all the essential developments. We hope this book can serve as a starting point for more case studies. Interested readers may benefit from learning about new concepts and techniques to model micro/nanoscale devices and systems and employing the techniques in their research. In the long term, we believe we will witness significant growth in the subject by integrating expertise from different fields.

The editors would like to express their sincere thanks to the National Science Foundation and Naval Research Office for providing financial support for the research efforts reported in this book.

MATLAB® is a registered trademark of The MathWorks, Inc. For product information, please contact:

The MathWorks, Inc.

3 Apple Hill Drive

Natick, MA 01760-2098 USA

Tel: 508 647 7000

Fax: 508-647-7001

E-mail: info@mathworks.com

Web: www.mathworks.com

The Editors

Ning Xi, DSc, is a University Distinguished Professor and John D. Ryder Professor of Electrical and Computer Engineering at Michigan State University, where he served previously as the director of the Robotics and Automation Laboratory. Currently, he is the head and chair professor of the Department of Mechanical and Biomedical Engineering at the City University of Hong Kong. He received his D.Sc. in systems science and mathematics from Washington University in St. Louis. He is the author of several award-winning papers in conference proceedings and recognized journals in his field.

Guangyong Li, PhD, is an associate professor in the Department of Electrical and Computer Engineering at the University of Pittsburgh. He received his PhD in electrical engineering from Michigan State University. He has published numerous papers in peer-reviewed journals and conference proceedings on nanorobotic manipulation, nanoscale characterization, and multiscale simulation of organic solar cells.

Mingjun Zhang, DSc, is an associate professor at the University of Tennessee, Knoxville. He received his DSc from Washington University in St. Louis, and his PhD from Zhejiang University, China. He was awarded the Young Investigator Program Award by ONR and the Early Career Award by the IEEE Robotics and Automation Society. Research results from his group have been published in *Proceedings of the National Academy of Sciences, Nano Letters, Advanced Functional Materials,* and *PLOS Computational Biology,* and highlighted by *Science, Nature,* and *AAAS Science Update.*

List of Contributors

Hongzhi Chen
Department of Electrical and
  Computer Engineering
Michigan State University
East Lansing, Michigan

Jennifer Y. Chen
Department of Chemistry
Drexel University
Philadelphia, Pennsylvania

Carmen Kar Man Fung
Hong Kong Productivity Council
Hong Kong

Xinghua Jia
Department of Mechanical,
  Aerospace and Biomedical
  Engineering
University of Tennessee
Knoxville, Tennessee

King Wai Chiu Lai
Department of Mechanical and
  Biomedical Engineering
City University of Hong Kong
Hong Kong

Scott C. Lenaghan
Department of Mechanical,
  Aerospace and Biomedical
  Engineering
University of Tennessee
Knoxville, Tennessee

Guangyong Li
Department of Electrical and
  Computer Engineering
University of Pittsburgh
Pittsburgh, Pennsylvania

Xiaobo Li
Department of Mechanical,
  Aerospace and Biomedical
  Engineering
University of Tennessee
Knoxville, Tennessee

Liming Liu
KLA-Tencor, Shanghai R&D
Shanghai, China

Stefan Oma Nwandu-Vincent
Department of Mechanical,
  Aerospace and Biomedical
  Engineering
University of Tennessee
Knoxville, Tennessee

Lynn S. Penn
Department of Chemistry
Drexel University
Philadelphia, Pennsylvania

Benjamin E. Reese
Department of Mechanical,
  Aerospace and Biomedical
  Engineering
University of Tennessee
Knoxville, Tennessee

Kristina Seiffert-Sinha
Department of Dermatology
University at Buffalo
and
Roswell Park Cancer Institute
Clinical and Translational Research
  Center
Buffalo, New York

Animesh A. Sinha
Department of Dermatology
University at Buffalo and Roswell
    Park Cancer Institute
Clinical and Translational Research
    Center
Buffalo, New York

Bo Song
Department of Electrical and
    Computer Engineering
Michigan State University
East Lansing, Michigan

Quan Tao
Department of Electrical and
    Computer Engineering
University of Pittsburgh
Pittsburgh, Pennsylvania

Tzyh-Jong Tarn
Center for Quantum Information
    Science and Technology
Beijing, China
and
Electrical and Systems Engineering
    Department
Washington University
St. Louis, Missouri

Yucai Wang
Department of Electrical and
    Computer Engineering
University of Pittsburgh
Pittsburgh, Pennsylvania

Fanan Wei
Department of Electrical and
    Computer Engineering
University of Pittsburgh
Pittsburgh, Pennsylvania

Rebing Wu
Department of Automation
Tsinghua University
and
Center for Quantum Information
    Science and Technology (TNLIST)
Beijing, China

Jun Xi
Department of Chemistry
Drexel University
Philadelphia, Pennsylvania

Ning Xi
Department of Electrical and
    Computer Engineering
Michigan State University
East Lansing, Michigan
and
Department of Mechanical and
    Biomedical Engineering
City University of Hong Kong
Hong Kong

Ruiguo Yang
Department of Electrical and
    Computer Engineering
Michigan State University
East Lansing, Michigan

Jing Zhang
Department of Automation,
Tsinghua University
and
Center for Quantum Information
    Science and Technology (TNLIST)
Beijing, China

Mingjun Zhang
Department of Mechanical,
    Aerospace and Biomedical
    Engineering
University of Tennessee
Knoxville, Tennessee

1

On the Principles of Quantum Control Theory

Re-Bing Wu, Jing Zhang, and Tzyh-Jong Tarn

CONTENTS

1.1 Introduction

In his 1959 famous lecture "Plenty of Room at the Bottom" [19], Richard Feynman considered the possibility of manipulating and controlling things at microscopic scales in the near future. Now his prediction has become reality with system assembly at the nanoscale level. When such systems are at a sufficiently low temperature, quantum effects will appear. From a cybernetical point of view, quantum control is a very useful theory for solving relevant problems in measurement, information transmittal, and state engineering.

What makes quantum control interesting are the nonclassical features of quantum mechanics. Unlike the classical world, the wave–particle duality sets no explicit boundary between particles and waves. An electromagnetic field can be quantized into particle-like photons, and wavelike properties such as diffraction can be observed with an atom or an electron. This makes it possible to manipulate waves like particles or vice versa. In other words, the essence of quantum control is to manipulate the coherence properties.

Early works on quantum control were motivated by control problems in plasma and laser devices, nuclear accelerators, and nuclear power plants [9–11]. In the 1980s, a serial study was casted to the modeling [46], controllability [24], invertibility [31], and quantum nondemolition filter problems [13]. Meanwhile, in Europe Belavkin started investigating filtering problems in quantum state estimation and feedback control applications [2–4]. Later, photonic control of chemical processes using lasers inspired a large number of quantum control studies that are still active today [1,36,37,43,49]. During this period, optimal control theory (OCT) was introduced to pulse shape design [33,45] based on known quantum Hamiltonians for the controlled molecules. In 1992, model-independent learning optimization algorithms [26] for real laboratory controls were proposed, which has led to over 150 successful experimental applications [7].

In the late 1990s, another tide of quantum control applications was developed by quantum information sciences. Since elementary quantum bits can be physically constructed by natural and artificial atoms (e.g., trapped ions, quantum dots, and superconducting circuits [8,22]), solving typical control problems, for example, state initialization, gate operation, error correction, and networking of quantum computers, is key to hardware implementation of quantum information processors.

The field of quantum control is still growing rapidly. Several monographs have been published on quantum control by authors from chemistry, quantum optics, and control theory [16,37,43,50], and on the state-of-art of quantum control from various review articles [7,17,47]. This chapter sketches fundamental concepts and problems in quantum control theory. In Section 1.2, physical mechanisms behind quantum control theory are discussed. Modeling and analysis problems are introduced in Section 1.3. In Section 1.4, open-loop and feedback control methods are briefly summarized. Finally, perspectives are presented.

1.2 Mechanism of Quantum Control

A nanoscale system can be modeled as either classical or quantum, although the former is nothing but an approximation of the latter under proper physical conditions. Mathematically, the quantum description is quantized from a classical model by replacing physical observables with operators that satisfy certain commutation relationships. In particular, to characterize the dynamics of quantum systems, it is essential to know the Hamiltonian that involves a potential energy function. As shown in Figure 1.1, the value of potential energy of a classical system can be an arbitrary real number. However, in quantum systems, only eigenvalues of the operator corresponding to the Hamiltonian can be recognized by a classical observer; they are called the energy levels. When the potential is a well, the energy levels are usually discrete, and the number of levels can be infinite when the well is infinitely

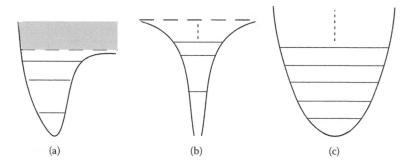

FIGURE 1.1
Spectrum with four different potentials: (a) potential well with finite depth, where both discrete (lower horizontal lines) and continuous levels (upper shaded area) exist; (b) an infinitely deep potential that is lower bounded; (c) an infinitely deep potential that is upper bounded.

deep (Figures 1.1b,c). The discretization of energy values is where we get the terminology "quantum," but it should be clarified that quantum energy levels are not always discrete, because a continuous spectrum (see Figure 1.1a) can exist when the energy is beyond the top of the potential.

These classically discernable levels form an orthogonal basis of the Hilbert space on which the quantum observables operate. Any physically admissible states can be expressed as a linear combination (with complex coefficients) of these basis vectors. In the so-called Schrödinger picture, the coefficients represent the probability amplitude of the corresponding states. For example, suppose a quantum system has only two energy levels, e_1 and e_2, then any physical allowable state (after normalization) can be expressed as

$$e = c_1 e_1 + c_2 e_2, \quad |c_1|^2 + |c_2|^2 = 1,$$

where $|c_{1,2}|^2$ represents the probability of the system staying at the state $e_{1,2}$. The phase difference between complex numbers c_1 and c_2 manifests the wavelike coherence feature of the quantum system.

Any physically measurable quantity can be represented by an operator $\hat{O}$ defined on the same Hilbert space, whose eigenvalues correspond to its recognizable values by a classical observer. Its expectation value under a quantum state ψ in the Hilbert space is calculated by the quadratic form $\langle \hat{O} \rangle = \psi^\dagger \hat{O} \psi$. These notations enable us to describe the famous Heisenberg uncertainty principle,

$$\langle \hat{A} \rangle \langle \hat{B} \rangle \geq \frac{1}{2} \langle [\hat{A}, \hat{B}] \rangle,$$

which reveals that two arbitrary noncommutative observables cannot be simultaneously precisely measured.

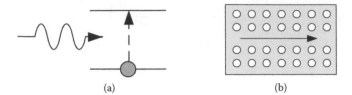

(a)   (b)

FIGURE 1.2
Schematics for quantum control scenarios based on field–matter interactions: (a) a two-level atom is controlled by a coherent field, and (b) a quantum field going through a bulk optimal medium is controlled by designed structures.

Control actions in quantum systems are based on the physical interaction between matter (e.g., atoms, ions, nuclear spins or quantum dots) and field (e.g., x-ray, light, or radio-frequency waves). Accordingly, the object to be controlled can be either a field or some matter. The control of matter is usually done using manipulable electromagnetic fields (see Figure 1.2a), while the control of field is done using a properly designed structure of the waveguide (e.g., photonic crystals, Figure 1.2b). It is also possible to use one field to control another field, but this has to be done with some linear or nonlinear medium because there is no direct interaction between electromagnetic fields.

For example, shaped coherent (laser) fields are widely used to manipulate quantum molecular dynamics. When the photon number contained in the field is large ($\sim 10^5$/cm^3 in each mode), the field amplitude can be approximated with high precision by its average value, which can be taken as a classical electromagnetic field. Moreover, when the size of the matter being radiated is far smaller than the wavelength of the light, the interaction with the field is only dependent on the position of the center of mass. Under properly chosen gauge, the interaction Hamiltonian can be written as

$$H_{\text{int}} = E(\mathbf{r}_0, t) \cdot \mathbf{D}, \tag{1.1}$$

where $\mathbf{D}$ is the dipole operator of the quantum system and the control field parameter $E(\cdot)$ is taken to be classical. When the size of the matter is comparable with the wavelength, a coupled Maxwell equation representing the evolution of the field will be required.

To simplify the analysis and design, it is always desirable to separate from the field–matter interacting system a minimal subsystem whose quantum effects cannot be ignored, and treat the rest as a classical system. Generally, a system has to be described by quantum mechanics if the fluctuation of the concerned observable is comparable with the gap between the eigenvalues of the observable. The following criteria are frequently used:

1. The control precision is comparable with the Heisenberg uncertainty.
2. An atom (natural or artificial) is taken as quantum when the environment temperature T is so low that the thermal fluctuation kT, where k is the Boltzmann constant, is smaller than the quantum energy gap δE.
3. An optical field is taken as quantum if it does not contain sufficiently many photons.

Following these rules, in the next section we introduce the general model of quantum control systems and an analysis of its controllability properties.

1.3 Modeling and Analysis of Quantum Control Systems

The evolution of the quantum state $\psi(t)$ is governed by the following Schrödinger equation:

$$i\hbar \frac{\partial \psi(t)}{\partial t} = \hat{H}(t)\psi(t), \tag{1.2}$$

where $\hat{H}(t)$ is the operator quantized from the classical Hamiltonian defined on the Hilbert space.

Without any classical approximation, the evolution of the joint state $\psi_{\text{tot}}(t)$ of the controller and the system can be written as

$$i\hbar \frac{\partial \psi_{\text{tot}}(t)}{\partial t} = \left\{ \hat{H}_0 + \sum_{k=1}^{r} \hat{u}_k(t)\hat{H}_k \right\} \psi_{\text{tot}}(t), \tag{1.3}$$

where the control $\hat{u}_k(t)$ is a time-dependent operator defined on the controller's Hilbert space. When the control can be treated as a classical variable, we can obtain the following bilinear quantum control system:

$$i\hbar \frac{\partial \psi(t)}{\partial t} = \left\{ H_0 + \sum_{k=1}^{r} u_k(t)H_k \right\} \psi(t), \tag{1.4}$$

where $\psi(t)$ is the quantum state of the system to be controlled. The operator H_0 is the internal Hamiltonian, and H_k is the control Hamiltonian via time-dependent control parameters $u_k(t)$. Here, the quantum system to be controlled can be either matter or field.

In quantum information sciences, the implementation of a quantum algorithm requires a directed evolution represented by the unitary operator $U(t)$ such that $\psi(t) = U(t)\psi(0)$ for any initial state $\psi(0)$. The control of $U(t)$ is described as follows:

$$i\hbar \frac{\partial U(t)}{\partial t} = \left\{ H_0 + \sum_{k=1}^{r} u_k(t)H_k \right\} U(t), \tag{1.5}$$

where $U(0)$ is always an identity operator (i.e., nothing is changed at the beginning of control).

Controllability is referred to as the ability to steer quantum systems between arbitrary states (or evolution operators) by varying the control in Equations 1.3, 1.4, or 1.5. From a practical point of view, it appears that controllability is not so important because not all state transitions are desired. However, a recent study [53] revealed that the loss of controllability may increase the complexity of finding optimal controls (i.e., search efforts)—the more the system is controllable, the easier is the search to reach a global optimal solution. Thus, it is warranted to delve further into the fundamental study of controllability.

First described in the early 1980s, the controllability criterion was proposed in terms of rank conditions of the Lie algebra generated by the system's free and control Hamiltonians. This was later extended to time-dependent systems [29] and systems with an infinite-dimensional controllability Lie algebra generated by internal and control Hamiltonians in Equation 1.4 [54]. Recently, a geometric analysis [6] showed that a two-level system interacting with a single optical mode is controllable over any finite-dimensional subspace without rotating-wave approximation, which is crucial for understanding the dynamics of strongly interacting quantum systems.

The controllability of infinite-dimensional systems, in particular those with unbounded or continuous spectrum, is extremely difficult because the set of states accessible from a fixed initial state is at the most dense in the Hilbert space. Therefore, the controllability will always have to be studied in a proper domain on which the functional analysis makes sense [24,54]. The results on a finite-dimensional case are much richer. For example, for molecular systems that can be approximated as finite dimensional, there are many studies on controllability with respect to the dipole structure [42] and degeneracies of the levels. With respect to the fully quantum model (1.3), an interesting result [38] shows that the state of a two-level system can be completely controlled by tuning the initial state of a coupled two-level system. Controllability of general systems can be enhanced by enriching the entanglement with the control source (or probe).

The previously mentioned studies assume that the quantum system to be controlled can be well isolated from the environment and can be precisely addressed by the control source. A typical example is the IVR (intramolecular vibrational resonance) that hindered the control of molecular reactions for quite a long time. Such uncontrollable environment interaction should be integrated into the model as a noise or a dissipation term [50], which may drive the system dynamics to its classical limit. Except when the control process can be completed within the decoherence time (e.g., in ultrafast control experiments), active controls should be posed against the decoherence effect. This is a fundamentally important control problem in quantum control theory [14,48,55].

1.4 Control Design Methodologies

The terminology "quantum control" is used rather loosely in the literature, as any attempt to choose a parameter can be called quantum control no matter what method is used. If the model can be precisely constructed and no significant disturbances exist, the open-loop control will be sufficient without having to measure the system output [7]. Otherwise, feedback control needs to be introduced for correcting the errors according to the measured result of the evolving state [17]. According to whether the controller and the plant are quantum or classical, one can classify the structure of the control system into the following four types:

1. *Open-loop control* (Figure 1.3a): The controller is treated as a classical system that has unidirectional causal action on the system.
2. *Direct coherent feedback control* (Figure 1.3b): The controller is treated as a quantum system with bidirectional causal relation with the system.
3. *Measurement-based feedback control* (Figure 1.3c): The controller is treated as a classical system that adjusts itself according to the classical information acquired from the measurement result of the system.
4. *Field-mediated coherent feedback control* (Figure 1.3d): The controller is treated as a quantum system with control loop closed by unidirectional field mediation.

1.4.1 Open-Loop Control of Quantum Systems

Open-loop control is the simplest and most broadly used control structure in the literature. Some of the open-loop control strategies were from physical intuition (e.g., dynamical decoupling method learned from the spin-echo phenomena in NMR systems [5,12]), and some were designed by more universal quantum optimal control theory and algorithms. In the following, we introduce the basic idea of quantum optimal control theory in the literature.

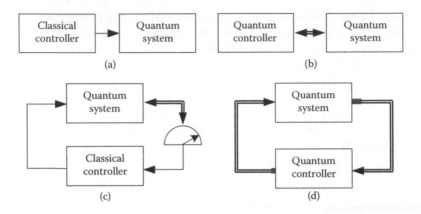

FIGURE 1.3
Types of quantum control: (a) open-loop control; (b) direct coherent feedback control; (c) measurement-based feedback control; and (d) field-mediated coherent feedback control.

Every target of control can be characterized by one or multiple costs as functional(s) of the control function, say $J[u(\cdot)]$. For example, the length of a chemical bond to be broken in chemical reactions can be taken as a cost to be maximized. Other cost functionals are, for example, the state transition probability [5], gate fidelity [32], coherence degree [55], fluence minimization [15], and the time consumed for control [27]. According to the Schrödinger equation (taken as a restriction on the control and state), one can derive the necessary condition for some $u(t)$ to minimize the cost functional if the control functions are not restricted. Otherwise, the generalized maximum principle [16] must be introduced to deal with the restrictions (e.g., limited field strength or bandwidth).

The optimal control is obtained by resolving the equations set by the necessary condition. When such equations (usually appearing in the form of nonlinear differential equations) are not analytically solvable, iterative algorithms can be designed to search the solution from an initial guess. From the control model Equation 1.4, it is possible to calculate the gradient of the cost with respect to the control function, following which a gradient-type algorithm [28,39,40] can be designed to steer the search of control solutions toward an optimal one. Iterative-type algorithms generally converge quickly, but can easily diverge if the initial guess was improperly chosen. By contrast, gradient-type algorithms are more stable as long as the step-size is sufficiently small, but the drawback is their convergence is very slow especially near the optimal solutions.

Quantum optimal control theory has been proven to be effective for finite-level or simple few-body systems [7]. However, the system Hamiltonians have to be precisely known, which is almost impossible in the laboratory. Even when the Hamiltonian is available, heavy computation will be not realistic on the numerical integration of the time-dependent Schrödinger

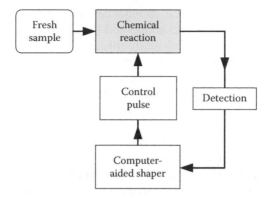

FIGURE 1.4
Schematic diagram of iterative learning control. The chemical reaction is controlled by a shaped laser pulse that is iteratively adjusted according to the measurement result. In each iteration, the control is performed on a fresh sample prepared at the same initial state.

equation in each iteration. To avoid this difficulty, evolutionary-type algorithms [26,44] were proposed (see Figure 1.4), where only the measurement output is needed for iterating the control function. This actually leaves the most difficult part of computation done by nature. Therefore, although the convergence may be much slower, learning algorithms are advantageous in the laboratory because not every detail of the real system must be known.

In quantum optimal control theory, all algorithms may suffer from the possible existence of false traps, that is, the search may converge to an undesired locally optimal solution that is not globally optimal. When the resource for optimization is not sufficient, the trapping behavior will become more severe. However, quick convergence was observed toward global optimal solutions in most reported simulations, and the reported experimental yields were remarkably improved as well [34]. This implies that traps are rare in practical control of quantum systems. Recently, this conjecture was supported via geometric analyses on the so-called quantum control landscape [34]. It was proven that if the system is controllable, then upon some mild assumptions there exist no obstacles on the landscape for maximization of observable expectation value [52], gate fidelity [35] in both closed and open systems [51]. Quantum control landscape theory has grown into a new research field, from which the details of the geometric landscape structure will provide greater understanding of quantum control and optimization so that more efficient algorithms can be developed.

1.4.2 Closed-Loop Control of Quantum Systems

In contrast to classical control theory, the body of literature on open-loop control is much larger than that of feedback control. This is not because feedback control is disadvantageous, but because the measurement required for feedback is much more difficult than what can be done with the control. The trade-off between inevitable backaction and the obtained information needs to be well understood. Nevertheless, quantum feedback control has already become a worldwide focus, as indicated by the 2012 Nobel Prize in physics awarded to Haroche and Wineland [23,41].

Measurement-based feedback control strategies can be further divided into two classes. The first is Markovian quantum feedback developed by H. M. Wiseman and G. J. Milburn [50], where quantum continuous weak measurement is adopted for feedback. With an idea learned from the quantum trajectory theory developed in quantum optics, the backaction of the measurement on the system is modeled as classical noise (either discontinuous jump noise or continuous Wiener white noise) that disturbs the system dynamics. The other approach is the Bayesian feedback control, initially proposed by A. C. Doherty [18], where the control strategy is determined based on the estimated state from the measurement output. The estimator is also called a "quantum filter" [2]. In principle, Bayesian feedback may outperform the Markovian feedback because more information extracted from the measurement history can be used to determine the control.

In addition to the backaction, the measurement-based feedback approaches are mainly limited by the time scale of quantum dynamics, which is often too fast to be followed by a classical controller. Regarding these difficulties, coherent feedback strategies have received much attention in recent years. As proposed by S. Lloyd [30], the simplest way to introduce coherent feedback is to couple directly the controlled quantum system to the quantum controller (Figure 1.3b). An alternative approach that is closer to measurement-based feedback control is field-mediated coherent feedback, where the quantum signal is carried by an intermediate field from the system to the controller (Figure 1.3d). This approach [21,25] was based on the quantum description of input–output theory developed by Gardiner et al. [20] in the 1980s. Compared with measurement-based feedback, the coherent feedback loop is more easily realized in experiments, thereby making it more suitable for scalable control circuits in future applications.

1.5 Perspectives

Quantum control is a developing field that is open to studies in various areas. Limited by fundamental experimental technologies for manipulating and observing quantum phenomena, experimental application of quantum control is still in its infancy except for the case of learning control. The fusion of control theory and quantum physics calls for collaborations between physicists and engineers. In the past decade, the gap between physicists and engineers has been greatly reduced, along with the rapidly developing experimental technologies and emerging common interests. There will be, undoubtedly, tremendous research opportunities for engineers in nanotechnology in the next decades.

Acknowledgments

The authors acknowledge support from NSFC (No.60904034, 61134008, 61174084) and the Tsinghua National Laboratory for Information Science and Technology (TNList) Cross-Discipline Foundation.

References

1. Bandrauk, A., Delfour, M., and Bris, C.L. (eds.). Quantum control: Mathematical and numerical challenges, *CRM Proceedings and Lecture Notes*, vol. 33. American Mathematical Society, Providence (2003).
2. Belavkin, V. Optimal quantum filtration of Markovian signals. *Problems of Information Transmission* **7**(5), 1–14 (1978).
3. Belavkin, V. and Grishanin, B. Optimum estimation in quantum channels by the generalized Heisenberg inequality method. *Problems of Information Transmission* **9**(3), 209–215 (1972).
4. Belavkin, V. and Grishanin, B. Optimum measurement of quantum variables. *Problems of Information Transmission* **8**(3), 259–265 (1972).
5. Bergmann, K., Theuer, H., and Shore, B.W. Coherent population transfer among quantum states of atoms and molecules. *Reviews of Modern Physics* **70**, 1003–1025 (1998).
6. Bloch, A., Brockett, R., and Rangan, C. Finite controllability of infinite-dimensional quantum systems. *IEEE Transactions on Automated Control* **55**(8), 1797–1805 (2010).
7. Brif, C., Chakrabarti, R., and Rabitz, H. Control of quantum phenomena: Past, present, and future. *New Journal of Physics* **12**, 075,008 (2010).
8. Buluta, I., Ashhab, S., and Nori, F. Natural and artificial atoms for quantum computation. *Reports on Progress in Physics* **74**(10), 104,401 (2011). DOI 10.1088/0034-4885/74/10/104401.
9. Butkovskii, A. and Pustyl'nikova, E. Controlling the coherent states of a quantum oscillator. *Automation and Remote Control* **43**, 1393–1398 (1982).
10. Butkovskii, A. and Pustyl'nikova, E. Control of coherent states of quantum systems with a quadratic Hamiltonian. *Automation and Remote Control* **45**, 1000–1008 (1984).
11. Butkovskii, A. and Samoilenko, Y. Control of quantum-mechanical processes and systems, *Mathematics and Its Applications*, vol. 56. Kluwer Academic Publishers, Boston (1990).

12. Chelkowski, S. and Bandrauk, A.D. Raman chirped adiabatic passage: A new method for selective excitation of high vibrational states. *Journal of Raman Spectroscopy* **28**(6), 459–466 (1997).
13. Clark, J., Ong, C., Tarn, T.J., and Huang, G.M. Quantum nondemolition filters. *Mathematical Systems Theory* **18**, 33–55 (1985).
14. Clausen, J., Bensky, G., and Kurizki, G. Bath-optimized minimal-energy protection of quantum operations from decoherence. *Physical Review Letters* **104**(4) (2010). DOI 10.1103/PhysRevLett.104.040401.
15. D'Alessandro, D. Optimal control of two level quantum systems. *IEEE Transactions on Automatic Control* **46**, 866–876 (2001).
16. D'Alessandro, D. *Introduction to Quantum Control and Dynamics*. Chapman & Hall/CRC, Boca Raton (2008).
17. Daoyi, D. and Petersen, I. Quantum control theory and applications: A survey. *IET Control Theory and Applications* **4**(12), 2651–2671 (2010).
18. Doherty, A.C., Habib, S., Jacobs, K., Mabuchi, H., and Tan, S.M. Quantum feedback control and classical control theory. *Physical Review A* **62**, 012,105 (2000).
19. Feynman, R. There's plenty of room at the bottom. Annual Meeting of the American Physical Society, California Institute of Technology, Pasadena, CA (December 29, 1959).
20. Gardiner, C.W. and Collett, M.J. Input and output in damped quantum systems: Quantum stochastic differential equations and the master equation. *Physical Review A* **31**, 3761–3774 (1985).
21. Gough, J. and James, M. The series product and its application to quantum feedforward and feedback networks. *IEEE Transactions on Automatic Control* **54**(11), 2530–2544 (2009).
22. Grigorenko, I. and Rabitz, H. Optimal control of the local electromagnetic response of nanostructured materials: Optimal detectors and quantum disguises. *Applied Physics Letters* **94**(25), 253107 (2009).
23. Hinds, E. and Blatt, R. Nobel 2012 physics: Manipulating individual quantum systems. *Nature* **492**(7427), 55 (2012).
24. Huang, G.M., Tarn, T.J., and Clark, J.W. On the controllability of quantum-mechanical systems. *Journal of Mathematical Physics* **24**, 2608–2618 (1983).
25. James, M., Nurdin, H., and Petersen, I. Control of linear quantum stochastic systems. *IEEE Transactions on Automatic Control* **53**(8), 1787–1803 (2008).
26. Judson, R. and Rabitz, H. Teaching lasers to control molecules. *Physical Review Letters* **68**, 1500 (1992).
27. Khaneja, N., Brockett, R., and Glaser, S. Time optimal control in spin systems. *Physical Review A* **63**(3), 032,308 (2001).
28. Khaneja, N., Reiss, T., Kehlet, C., Schulte-Herbrüggen, T., and Glaser, S.J. Optimal control of coupled spin dynamics: Design of NMR pulse sequences by gradient ascent algorithms. *Journal of Magnetic Resonance* **172**, 296–305 (2005).
29. Lan, C., Tarn, T.J., Chi, Q.S., and Clark, J.W. Analytic controllability of time-dependent quantum control systems. *Journal of Mathematical Physics* **46**(5), 052102 (2005).
30. Lloyd, S. Coherent quantum feedback. *Physical Review A* **62**, 022,108 (2000).
31. Ong, C., Huang, G., Tarn, T., and Clark, J. Invertibility of quantum-mechanical control systems. *Mathematical Systems Theory* **17**(4), 335–350 (1984).

32. Palao, J.P. and Kosloff, R. Quantum computing by an optimal control algorithm for unitary transformations. *Physical Review Letters* **89**, 188,301 (2002).
33. Peirce, A., Dahleh, M., and Rabitz, H. Optimal control of quantum-mechanical systems: Existence, numerical approximation, and applications. *Physical Review A* **37**, 4950–4964 (1988).
34. Rabitz, H., Hsieh, M., and Rosenthal, C. Quantum optimally controlled transition landscapes. *Science* **303**(5666), 1998–2001 (2004).
35. Rabitz, H., Hsieh, M., and Rosenthal, C. The landscape for optimal control of quantum-mechanical unitary transformations. *Physical Review A* **72**, 52,337 (2005).
36. Rabitz, H., de Vivie-Riedle, R., Motzkus, M., and Kompa, K. Whither the future of controlling quantum pheonomena? *Science* **288**, 824–828 (2000).
37. Rice, S. and Zhao, M. *Optical Control of Molecular Dynamics*. John Wiley & Sons, New York (2000).
38. Romano, R. and D'Alessandro, D. Environment-mediated control of a quantum system. *Physical Review Letters* **97**, 080,402 (2006).
39. Rothman, A., Ho, T.S., and Rabitz, H. Observable-preserving control of quantum dynamics over a family of related systems. *Physical Review A* **72**(2), 023416 (2005).
40. Rothman, A., Ho, T.S., and Rabitz, H. Quantum observable homotopy tracking control. *Journal of Chemical Physics* **123**(13), 134104 (2005).
41. Sayrin, C., Dotsenko, I., Zhou, X., Peaudecerf, B., Rybarczyk, T., Gleyzes, S., Rouchon, P., Mirrahimi, M., Amini, H., Brune, M., Raimond, J.M., and Haroche, S. Real-time quantum feedback prepares and stabilizes photon number states. *Nature* **477**(7362), 73–77 (2011).
42. Schirmer, S.G., Fu, H., and Solomon, A.I. Complete controllability of quantum systems. *Physical Review A* **63**(6), 063,410 (2001).
43. Shapiro, M. and Brumer, P. *Principles of the Quantum Control of Molecular Processes*. Wiley-Interscience, Hoboken (2003).
44. Shir, O., Beltrani, V., Bäck, Th., Rabitz, H., and Vrakking, M. On the diversity of multiple optimal controls for quantum systems. *Journal of Physics B* **41**(7), 074,021 (2008).
45. Tannor, D., Kosloff, R., and Rice, S. Coherent pulse sequence induced control of selectivity of reactions: Exact quantum mechanical calculations. *Journal of Chemical Physics* **85**(10), 5805–5820 (1986).
46. Tarn, T.J., Huang, G.M., and Clark, J. Modelling of quantum mechanical control systems. *Mathematical Modelling* **1**, 109–121 (1980).
47. Vandersypen, L. and Chuang, I. NMR techniques for quantum control and computation. *Reviews of Modern Physics* **76**(4), 1037–1069 (2004).
48. Viola, L. and Lloyd, S. Dynamical suppression of decoherence in two-state quantum systems. *Physical Review A* **58**(4), 2733–2744 (1998). DOI 10.1103/PhysRevA.58.2733.
49. Warren, W., Rabitz, H., and Dahleh, M. Coherent control of quantum dynamics: The dream is alive. *Science* **259**, 1581–1588 (1993).
50. Wiseman, H. *Quantum Measurement and Control*. Cambridge University Press, Cambridge (2010).

51. Wu, R., Pechen, A., Rabitz, H., Hsieh, M., and Tsou, B. Control landscapes for observable preparation with open quantum systems. *Journal of Mathematical Physics* **49**, 022,108 (2008).
52. Wu, R., Rabitz, H., and Hsieh, M. Characterization of the critical submanifolds in quantum ensemble control landscapes. *Journal of Physics A* **41**, 015,006 (2008).
53. Wu, R.B., Hsieh, M.A., and Rabitz, H. Role of controllability in optimizing quantum dynamics. *Physical Review A* **83**, 062,306 (2011).
54. Wu, R.B., Tarn, T.J., and Li, C.W. Smooth controllability of infinite-dimensional quantum-mechanical systems. *Physical Review A* **73**, 012,719 (2006).
55. Zhang, J., Li, C.W., Wu, R.B., Tarn, T.J., and Liu, X.S. Maximal suppression of decoherence in Markovian quantum systems. *Journal of Physics A: Mathematical and General* **38**(29), 6587 (2005).

2

Modeling and Simulation of Silicon Nanowire–Based Biosensors

Guangyong Li, Yucai Wang, and Quan Tao

CONTENTS

2.1 Introduction

Quasi-one-dimensional semiconducting nanowires are considered the best candidates among transducer elements for biosensing because of their appealing characteristics such as high sensitivity due to the quantum confinement and the large surface-to-volume ratio, high stability due to the crystal structure, and potential to be configured as field-effect transistors (FETs). Several pioneering studies have demonstrated the success of direct electrical detection of biological macromolecules using carbon nanotubes [1–3], semiconducting nanowires [4–9], and conducting polymer nanofilaments [10,11]. Among these biosensors, silicon nanowire (SiNW)-based biosensors (Figure 2.1) are considered promising label-free biomolecule detectors because of their compatibility with microelectronics and their great stability in liquid condition. Label-free SiNW-based biosensors configured as FETs can detect the target/receptor binding events that affect the local chemical potential on the surface of nanowires. The local potential on the surface of nanowires effectively works as a "gating" voltage that modulates the source (S) to drain (D) current.

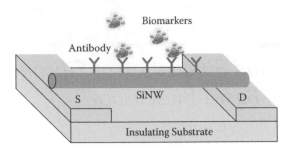

FIGURE 2.1
SiNW biosensor configured as an FET, where the gating voltage comes from the biomarkers.

Despite advances in experimental study, the underlying detection mechanism of SiNW-based FET biosensing is not well understood, and the path for future optimization in sensitivity needs to be elaborated. Simulation studies will not only help us understand the physics of biosensing using SiNW-based FETs but also provide guidance for optimized design of the devices such as the size of the nanowires, the dimension of the devices, the packing density of antibodies, and the concentration of the electrolyte.

In this chapter, we presents a comprehensive simulation study on single SiNW-based FET biosensors for detecting biotin/streptavidin binding [12,13]. The biotin/streptavidin system exhibits the strongest non-covalent biological interaction known and is widely demonstrated as a model system to study bio-recognition between proteins and other biomolecules [14]. Thus, understanding the detection mechanism of biotin/streptavidin binding using SiNW-based FET biosensors is of special interest since this can serve as a stepping stone to improving the sensitivity of SiNW-based FET biosensors.

2.2 The Basics of SiNW-Based FET Biosensors

The general schematic of a SiNW FET biosensor is shown in Figure 2.2a. A SiNW core is placed on a silicon dioxide substrate with its two ends connected to two electrodes (source and drain). The SiNW core is enveloped by a silicon oxide layer whose surface is functionalized with specific receptor molecules (Y), which can recognize and bind only to the target molecules (flower). The SiNW is immersed in the electrolyte, which contains target molecules as well as positively charged (e.g., Na^+) and negatively charged (e.g., Cl^-) ions. A back gate electrode (U_{BG}) and a reference electrode (U_G) are connected to the system to adjust the bulk electrolyte potential. The electrolyte is used as a buffer solution that resists pH change upon the addition of a small amount of acid or base, or upon dilution. The target molecules diffuse through the solution and reach the SiNW, are captured by the receptor molecules, and finally bind

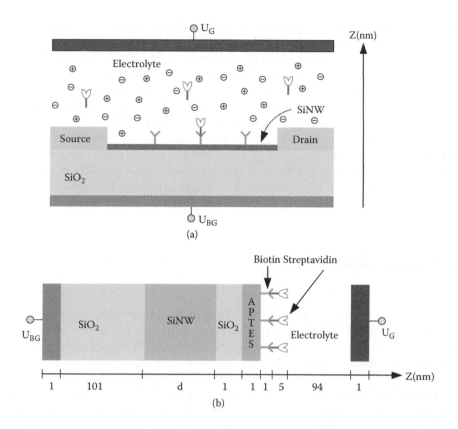

FIGURE 2.2
(a) General schematic of SiNW FET biosensor for detection of biotin/streptavidin binding. The surface of SiNW is functionalized with receptors (Y) for target biomolecules (flower). (b) Cross-section of SiNW FET biosensor for detection of biotin/streptavidin binding between the source and drain region along the z axis. On the surface of SiNWs there is a native oxide that is covalently bound to a layer of APTES molecules, and biotin molecules are chemically attached to the APTES surface. The biosensor is immersed in aqueous solution to detect streptavidin in the solution. The numbers denote the dimensions of various layers along the z axis.

the receptor molecules on the silicon oxide surface. Many biomolecules carry charges under normal physiological conditions. For example, DNA carries negative charges, whereas the net charges of protein depend on the pH value of the electrolyte. The charges of the bound target molecules interact with the charge in the SiNW and modulate the conductance of the SiNW. The changes of the SiNW conductance can be used to measure the concentration of the target molecules in the buffer solution, and thus derive the original concentration of the target molecules in the source sample.

Biotin molecules cannot be directly immobilized on the silicon oxide surface. A layer of aminopropyltriethoxysilane (APTES) molecules is covalently bound to the silicon oxide surface, and biotin molecules (receptor molecules) are chemically attached to the APTES surface. Streptavidin molecules in aqueous solution can be specifically bound to biotin molecules. Figure 2.2b shows the cross-section view of the SiNW FET biosensor along the z axis. Here, we assume the thickness of the silicon oxide substrate is about 100 nm, the thickness of the silicon oxide layer enveloping the SiNW core is 1 nm, and the thickness of the APTES molecule layer is 1 nm. The dimensions of the biotin molecule and the streptavidin molecule are 0.52 nm × 1.00 nm × 2.10 nm and 4.5 nm × 4.5 nm × 5.0 nm, respectively. So we model the biotin and streptavidin molecules as spheres and assume the thickness of biotin layer is 1 nm and the thickness of streptavidin is 5 nm. The electrolyte (NaCl solution with various concentrations; pH value is assumed as 7) region is 100 nm, and the thickness of metal electrode is 1 nm. The length, diameter (denoted as d) and the doping density (assuming a p-type doping) of the SiNW are variable.

2.3 Theoretical Approaches

To detect the biotin/streptavidin binding in the electrolyte, generally two states of the SiNW FET biosensor should be simulated. In the first state, only the receptor molecules (biotin) are attached to the APTES layer surface. In the second state, the target biomolecules (streptavidin) are attached to the biotin surface due to specific binding. The change of charge distribution arising from the biotin/streptavidin binding modulates the electrostatic potential on the silicon oxide surface of the SiNW, and hence modulates charge distribution inside the SiNW core. As a consequence, the current flows in the SiNW will be changed accordingly. The I-V characteristics of the SiNW can be examined to determine the sensitivity of the biosensor.

Using the drift-diffusion charge transport model, the carrier transport in the SiNW core can be written as [15]

$$J_{e,h} = q\mu_{e,h}n_{e,h}\nabla\Phi \pm qD_{e,h}\nabla n_{e,h} \tag{2.1}$$

where q is elementary charge, J is the current density, μ is the mobility of the carriers, n is the carrier density, Φ is the electric potential, and D is the diffusion coefficient. The subscripts e and h denote electron and hole, respectively. The relationship between the electrostatic potential and the carrier concentrations can be described using Poisson's equation,

$$-\nabla\left(\varepsilon_{Si}\nabla\Phi(r)\right) = q(n_h - n_e + N_D - N_A) \tag{2.2}$$

where ε_{Si} is the dielectric constant of silicon, r is the spatial coordinate, and N_D and N_A are the donor and acceptor concentrations within the SiNW, respectively. Here we assume the dopants are completely ionized.

In the silicon oxide layer, we assume that there is no defect in the native oxide and neglect any interface traps and fixed oxide charges. The electrostatic potential in the silicon oxide is given by Poisson's equation,

$$-\nabla\left(\varepsilon_{Ox}\nabla\Phi(r)\right)=0 \tag{2.3}$$

where ε_{ox} is the dielectric constant of silicon oxide.

The APTES molecule layer serves as a passivation layer, and no charge carriers are assumed to be present within this layer. Similarly, the electrostatic potential in the APTES molecule layer is given by Poisson's equation,

$$-\nabla\left(\varepsilon_{APTES}\nabla\Phi(r)\right)=0 \tag{2.4}$$

where ε_{APTES} is the dielectric constant of APTES molecule layer.

In SiNW FETs, the gate voltage is externally applied via a gate electrode enveloping the silicon oxide layer. In SiNW FET biosensors, no gate electrode is present, and the gate voltage is determined by solving the nonlinear Poisson–Boltzman equation (NPBE) [16],

$$-\nabla\left(\varepsilon_r(r)\nabla\Phi(r)\right)=\rho_{fixed}(r)+\rho_{mobile}(r) \tag{2.5}$$

where ρ_{fixed} is the charge density in the biotin layer and the bounded streptavidin molecules, and ρ_{mobile} is the charge density of mobile ions in the electrolyte, which accounts for the Debye screening effect. The dielectric constant of electrolyte is shown as ε_r. The fixed charge density ρ_{fixed} can be modeled using the delta function,

$$\rho_{fixed}=q\sum_{i}^{N}z_i\delta(r-r_i) \tag{2.6}$$

where N is the number of fixed charges present, z_i and r_i are their charges and positions, and $\delta(r)$ is the Dirac delta function. The distribution of mobile charge density ρ_{mobile} can be described using the Boltzmann distribution,

$$\rho_{mobile}=\sum_{j}^{N_{ions}}qz_jn_\infty\exp\left(\frac{-qz_j\Phi}{K_BT}\right) \tag{2.7}$$

where N_{ions} is the number of the mobile ions, z_j is the charge of the ion, n_∞ is the concentration of the ions at a distance of infinity from the solute, and Φ is the electrical potential. K_B is Boltzmann's constant, and T is the absolute temperature.

For the special case of a 1–1 electrolyte (e.g., Na⁺–Cl⁻) with bulk concentration n, ρ_{mobile} can be written as [17]

$$\rho_{mobile} = -\frac{K_B T}{q}\varepsilon_r \kappa^2 \sinh\left(\frac{q\Phi}{K_B T}\right) \tag{2.8}$$

where

$$\kappa = \sqrt{\frac{2n_\infty q^2}{K_B T \varepsilon_r \varepsilon_0}} \tag{2.9}$$

is the Debye–Hückel screening parameter. κ^{-1} is called the Debye screening length, an important parameter to characterize the thickness of the electrical double layer. For the special case of a 1–1 electrolyte (e.g., Na⁺–Cl⁻), Debye length versus electrolyte concentration is illustrated in Figure 2.3 and listed in Table 2.1. The SiNW can sense only charged molecules within the Debye screening length. The charged molecules beyond the Debye screening length will be screened by the ions in the electrolyte. As can be seen, NaCl concentrations larger than 10 mM are not suitable for detection of the biotin/streptavidin binding because part of the charge on the biomolecules will be screened (Figure 2.4). The lines in Figure 2.4 show the different Debye screening lengths (from SiNW to the lines, not scaled here) according to different NaCl concentrations. Apparently, higher NaCl concentrations result in shorter Debye screening lengths.

Physically, the electrostatic potential function $\Phi(r)$ should be continuous at the interface of the various layers. Also, for the infinite domain, the

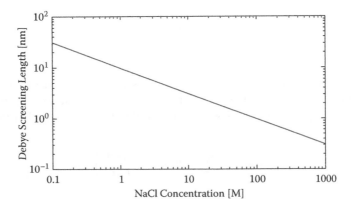

FIGURE 2.3
Debye screening lengths versus different NaCl concentrations.

TABLE 2.1

Debye Screening Lengths versus Different
NaCl Concentrations

NaCl Concentration	Debye Screening Length (nm)
0.1 mM	30.79
1 mM	9.74
10 mM	3.08
100 mM	0.97
1 M	0.31

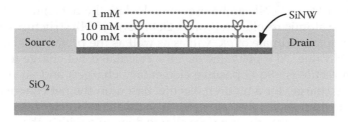

FIGURE 2.4
Debye screening scheme (the distance is not to scale).

electrostatic potential at infinity should be zero, i.e., $\Phi(\infty) = 0$. As long as we find the solution for the NPBE of Equation 2.5, the electrostatic potential in other layers can be solved by applying the continuous condition. And finally we can get the current density inside the SiNW using Equation 2.1.

The protein carries a net charge if the pH value of the electrolyte is not equal to the isoelectric point (pI) value of the protein. The protein net charge at a given pH value can be estimated using the Henderson–Hasselbalch equation [18].

$$pH = pK_a + \log_{10}\frac{[A^-]}{[HA]} = pK_a + \log_{10}\frac{\alpha}{1-\alpha} \tag{2.10}$$

where α is the fractional dissociation of any ionizable group [18] and pK_a is the acid dissociation constant. In the charge determination procedure [19], the pK_a for each type of ionizable group is assigned a magnitude. Knowledge of the amino acid sequence (or composition) of a protein is then used to calculate the net charge. The protein net charge arising from n_i ionizable groups of type i, $(Z_p)_i$, is given by [18]

$$(Z_p)_i = n_i z_i \alpha_i \tag{2.11}$$

if the ionizable group, i, is anionic (z_i is negative), or

$$(Z_p)_i = n_i z_i (1 - \alpha_i) \tag{2.12}$$

if the ionizable group, i, is cationic (z_i is positive). The pK_a value of biotin is 4.66 [20], so at pH = 7, each biotin molecule carries a net charge of $\alpha = 1/(1+10^{4.66-7})$ $\approx -1q$. This result coincides with the experimental result in Reference 21.

The net charge of streptavidin is contributed by terminal amino acids and charged amino acid side-chains within the protein sequence [22]. It can be determined by computer programs given the protein sequence data and a set of pK_a values for proton dissociation from ionizable groups [19]. A web-based server H++ [23,24] can be used to calculate the net charge of strepta-vidin. Given atomic resolution Protein Data Bank (PDB) structures, H++ can perform a quick estimate of pK_a values as well as the protein net charge at a specific pH value. At a pH value of 7, the calculated net charge for a strep-tavidin molecule is +9q (9 positive elementary charges) and −q (1 negative elementary charge) for a biotin molecule. Based on the net charge of biotin and streptavidin molecules, along with the dimension information of the SiNW, we can estimate the interface charge density ρ within the biotin layer before and after the binding of streptavidin molecules.

2.4 Simulation Results and Discussions

2.4.1 Surface Potential on SiNW

In general, the NPBE of Equation 2.5 can be solved using only computational methods due to the rapid exponential nonlinearities, discontinuous coefficients, delta functions, and infinite domain. And the accuracy and stability of the solution to the NPBE are quite sensitive to the boundary layer between the electrolyte and the biomolecules, which defines the solvent-accessible surface. In this work, we employ a nanodevice simulator nextnano³ [25] to solve the NPBE and to find the electrostatic potential values in the electro-lyte, APTES molecule layer, and silicon oxide layer. The nextnano³ uses the Gouy–Chapman theory [17] to solve the NPBE for a one-dimension planar ion-selective field-effect transistor (ISFET) biosensor [26].

There are several key assumptions in the Gouy–Chapman theory. One is that the surface of the ISFET biosensor is infinite and planar, so that the inter-face between the biomolecules and the electrolyte can be described as an infinite plane that separates two regions, each with a different dielectric con-stant. Another one is that the surface of the ISFET biosensor is assumed to contain a constant fixed charge density ρ, with unit of charge per area. Under these assumptions, the electrostatic potential in the electrolyte is dependent on only the distance z from the membrane surface. For our SiNW FET biosensor, the electrostatic potential is dependent on only the distance r from the APTES molecules layer along the radius direction of the SiNW. Thus, it is feasible to use nextnano^3 to solve the NPBE for our SiNW FET biosensor. And the critical step is to calculate the interface charge density ρ, that is, the charge density in the biotin layer before and after the binding of streptavidin molecules.

Nextnano^3 needs the following parameters to calculate the electrostatic potential in the electrolyte, APTES molecules layer, and silicon oxide layer: the dimension information of various layers and the electrolyte, the doping density of the SiNW, the ion (NaCl) concentration of the electrolyte, the interface charge density ρ, and the voltages applied on electrodes U_{BG} and U_G. The dimension information on various layers and electrolyte is shown in Figure 2.2b. The interface charge density ρ can be calculated using the dimension information and the net charge of biotin/streptavidin. The voltages applied on electrodes U_{BG} and U_G are set to zero. Now we are ready to calculate the electrostatic potential corresponding to different SiNW diameters, doping densities, and ions concentrations of the electrolyte.

The APTES surface potential change $\Delta\Phi_s$ is defined as

$$\Delta\Phi_s = \Phi_s(n) - \Phi_s^{ref}, \qquad (2.13)$$

where n denotes the number of streptavidin molecules bound to the avidin molecules, and Φ_s^{ref} is the reference voltage when no streptavidin molecules are bound to the avidin molecules. For a SiNW with diameter $d = 10$ nm, doping density $p = 10^{15} cm^{-3}$, length $L = 50$ nm, and in the electrolyte with NaCl concentration 10 mM, Φ_s^{ref} is about 0.4V.

Assuming the number of streptavidin molecules bound to the SiNW surface varies from 0 to 10, the calculated surface potential change $\Delta\Phi s$ is shown in Figure 2.5. As can be seen, the surface potential at the silicon oxide layer increases as the number of the bound streptavidin molecules increases. The curve is nonlinear due to the Debye screening effect of the electrolyte where the ions (Na$^+$, Cl$^-$) in the electrolyte screen the charges on the biotin and streptavidin molecules.

2.4.2 I-V Characteristics of SiNW FET Biosensors

The I-V characteristics of the SiNW FET biosensor can be obtained by solving the Poisson equation and the drift-diffusion equation using a MuGFET simulator [27]. MuGFET is a simulation tool for nanoscale multigate FET structures and SiNW FETs. It provides self-consistent solutions to the Poisson and drift-diffusion equation via a user-friendly graphical user interface.

We assume the source and drain extension of the SiNW is 10 nm, and the source and drain are n-type doped with doping concentration 10^{18} cm^{-3}. The source electrode is grounded. In our simulation for the silicon oxide

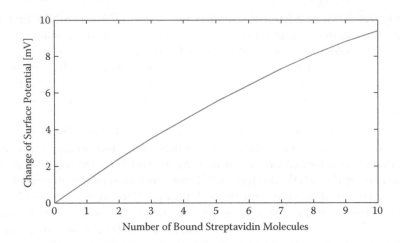

FIGURE 2.5
Change of silicon oxide surface potential versus number of streptavidin molecules bound.

surface voltage change, the reference voltage on the silicon oxide surface (the initial gate voltage for the SiNW FET) is around 0.4V. Binding of the streptavidin molecules leads to the gate voltage change of several millivolts.

It is important to find a proper drain voltage to make the SiNW FET work at a point that is very sensitive to the gate voltage change, that is, a small change in the gate voltage will lead to a big change in the drain current so that the drain current change is appreciable.

We calculate the I-V characteristics for three different gate voltages (Vg = 0.40V, 0.41V, 0.42V; these values are chosen because the reference voltage is about 0.4 V) while keeping the following parameters constant: SiNW diameter d = 10 nm, channel doping density p = 10^{15} cm^{-3}. The calculated I-V characteristic is shown in Figure 2.6. As can be seen, when the drain voltage is small, the SiNW FET works in the linear region and the drain current change for different gate voltages is very small. When the drain voltage is large enough, the SiNW FET works in the saturation region and the drain current change for the different gate voltage is distinct. Thus, the SiNW FET should work in the saturation region to have a good sensitivity for the gate voltage change. In the remaining simulations, we will fix the drain voltage as 1 V to make sure the SiNW FET work in the saturation region.

2.4.3 Sensitivity Analysis

We define the drain current change for different silicon oxide surface potentials (gate voltages) as

$$\Delta I_d = \frac{I_d(n) - I_d^{ref}}{I_d^{ref}} \times 100\% \qquad (2.14)$$

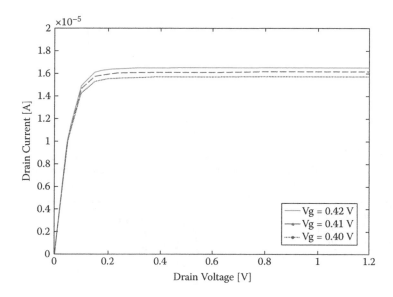

FIGURE 2.6
The I-V characteristics for different gate voltages.

where n denotes the number of streptavidin molecules bound to the avidin molecules and I_d^{ref} is the reference drain current when no streptavidin molecules are bound to the avidin molecules. The drain current change can be regarded as the sensitivity of the SiNW FET biosensor.

The changes of the drain current according to the number of bound streptavidin molecules in different NaCl concentrations are shown in Figure 2.7a. As can be seen, a lower concentration of NaCl will lead to a large change of silicon oxide surface potential and consequently a larger drain current change. Thus, we can conclude that lower NaCl concentrations will result in higher sensitivity of the SiNW FET biosensor.

Using a similar procedure, we investigate the influence of SiNW length to the sensitivity of the SiNW FET biosensor while keeping the following parameters constant: SiNW diameter $d = 10$ nm, doping density $p = 10^{15}$ cm^{-3}, and the NaCl concentration in the electrolyte is 10 mM. Three different SiNW lengths ($L = 50$ nm, 70 nm, and 90 nm) are investigated, and the results are shown in Figure 2.7b. As can be seen, SiNW with a shorter length has higher sensitivity. This is reasonable since for a shorter SiNW the traveling speed of carriers is faster, and there will be less carrier recombination.

The influence of SiNW diameter on the sensitivity of the SiNW FET biosensor is investigated while keeping the following parameters constant: SiNW length $L = 50$ nm, doping density $p = 10^{15}$ cm^{-3}, and the NaCl concentration in the electrolyte is 10 mM. Three different SiNW diameters ($d = 10$ nm, 12 nm, and 14 nm) are investigated, and the results are shown in Figure 2.7c. As can be seen, SiNW with a smaller diameter shows higher sensitivity. This is

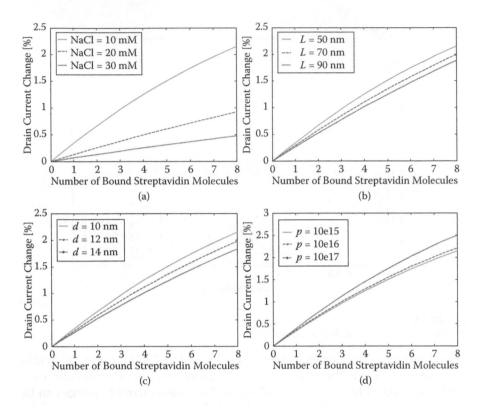

FIGURE 2.7
Drain current change versus number of streptavidin molecules bound for different NaCl concentrations: (a), nanowire length, (b), nanowire diameter, (c), and doping level (d).

justifiable since for thinner SiNW, the gate voltage has better control over the conducting channel inside the SiNW.

The influence of doping levels on the sensitivity of the SiNW FET biosensor is investigated while keeping the following parameters constant: SiNW diameter $d = 10$ nm, length $L = 50$ nm, and the NaCl concentration in the electrolyte is 10 mM. Three different doping levels ($p = 10^{15}$ cm^{-3}, 10^{16} cm^{-3}, and 10^{17} cm^{-3}) are investigated, and the results are shown in Figure 2.7d. As can be seen, SiNW with higher doping levels shows higher sensitivity.

2.4.4 Discussion

From the above analysis, we can conclude that SiNW with a shorter length, smaller diameter, higher doping level, and with lower NaCl concentration in the electrolyte has higher sensitivity. When these parameters are properly chosen, it is possible to achieve single streptavidin molecule detection. The following parameters are chosen for a SiNW FET biosensor: SiNW diameter $d = 10$ nm, length $L = 50$ nm, doping level $p = 10^{15}$ cm^{-3}, and the NaCl

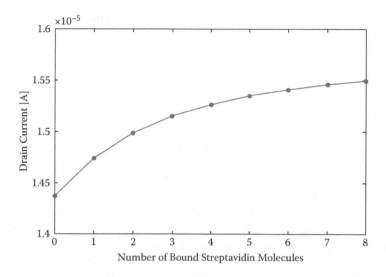

FIGURE 2.8
Drain current versus number of streptavidin molecules bound.

concentration in the electrolyte is 1 mM. With current technology, SiNW FET biosensors with these parameters can be readily fabricated. The drain current according to the number of bound streptavidin molecules is shown in Figure 2.8. When the number of the bound streptavidin molecules changes from 0 to 1, the drain current changes by 2.59% and since this change is in measurable range, recognition of single streptavidin molecules is feasible. Actually, the sensitivity can be even higher when a higher doping level is chosen.

2.5 Conclusions and Perspectives

In this chapter, using the biotin/streptavidin system as a model system, comprehensive modeling and simulation studies are presented to reveal the underlying detection mechanism of biotin/streptavidin binding using the SiNW FET biosensor. The detailed structures, modeling procedure, and simulation methods of the SiNW FET biosensor are presented. The simulation results are analyzed and the influence of parameters such as the dimension of the SiNW (diameter and length), the doping level of the SiNW, and surrounding environment (the ion concentration in the solvent) are investigated for the sensitivity of the SiNW FET biosensor. The detection limit of biotin/streptavidin binding using the SiNW FET biosensor, that is, the detection of single streptavidin molecule binding, is also investigated. The preliminary simulation results indicate that optimal sensor performance can be ensured by careful optimization of its parameters, and it is feasible to detect binding of single streptavidin molecules when optimal parameters are chosen. However, there is room for the sensitivity to be further improved. The modeling procedure and simulation methods shown in this work can be readily modified and adopted for detection of other receptor–target biomolecules pairs, such as ssDNA–dsDNA, antibody–protein, and antibody–virus, using SiNW FET biosensors. The simulation results can serve as a guideline for the design and optimization of SiNW FET biosensors in future studies.

Several problems need to be addressed in future studies. First, the Gouy–Chapman theory is suitable for solving the NPBE for a one-dimension planar ISFET biosensor. For this particular application, accuracy is good. However, it is an approximation to solve the NPBE for a SiNW FET biosensor using the Gouy–Chapman theory, but the accuracy cannot be guaranteed. Second, biotin and streptavidin molecules are modeled as spheres in this study, differing from their original structures and affecting simulation accuracy. In future studies, biotin and streptavidin should be modeled using their exact structure. A 3D NPBE solver must be used to solve the surface potential for the SiNW FET biosensor.

References

1. K. Besteman, J.-O. Lee, F. G. M. Wiertz, H. A. Heering, and C. Dekker, "Enzyme-coated carbon nanotubes as single-molecule biosensors," *Nano Letters,* vol. 3, no. 6, pp. 727–730, 2003.
2. A. Star, E. Tu, J. Niemann, J.-C. P. Gabriel, C. S. Joiner, and C. Valcke, "Label-free detection of DNA hybridization using carbon nanotube network field-effect transistors," *Proceedings of the National Academy of Sciences of the United States of America,* vol. 103, no. 4, pp. 921–926, 2006.
3. K. Maehashi, T. Katsura, K. Kerman, Y. Takamura, K. Matsumoto, and E. Tamiya, "Label-free protein biosensor based on aptamer-modified carbon nanotube field-effect transistors," *Analytical Chemistry,* vol. 79, no. 2, pp. 782–787, 2006.
4. Y. Cui, Q. Wei, H. Park, and C. M. Lieber, "Nanowire nanosensors for highly sensitive and selective detection of biological and chemical species," *Science,* vol. 293, no. 5533, pp. 1289–1292, 2001.
5. J.-i. Hahm, and C. M. Lieber, "Direct ultrasensitive electrical detection of DNA and DNA sequence variations using nanowire nanosensors," *Nano Letters,* vol. 4, no. 1, pp. 51–54, 2004.
6. F. Patolsky, G. Zheng, O. Hayden, M. Lakadamyali, X. Zhuang, and C. M. Lieber, "Electrical detection of single viruses," *Proceedings of the National Academy of Sciences of the United States of America,* vol. 101, no. 39, pp. 14017–14022, 2004.

7. Z. Li, Y. Chen, X. Li, T. I. Kamins, K. Nauka, and R. S. Williams, "Sequence-specific label-free DNA sensors based on silicon nanowires," *Nano Letters,* vol. 4, no. 2, pp. 245–247, 2004.
8. J.-i. Hahm, and C. M. Lieber, "Direct ultrasensitive electrical detection of DNA and DNA sequence variations using nanowire nanosensors," *Nano Letters,* vol. 4, no. 1, pp. 51–54, 2003.
9. T. Cohen-Karni, B. P. Timko, L. E. Weiss, and C. M. Lieber, "Flexible electrical recording from cells using nanowire transistor arrays," *Proceedings of the National Academy of Sciences,* vol. 106, no. 18, pp. 7309–7313, 2009.
10. K. Ramanathan, M. A. Bangar, M. Yun, W. Chen, N. V. Myung, and A. Mulchandani, "Bioaffinity sensing using biologically functionalized conducting-polymer nanowire," *Journal of the American Chemical Society,* vol. 127, no. 2, pp. 496–497, 2004.
11. N. Chartuprayoon, C. M. Hangarter, Y. Rheem, H. Jung, and N. V. Myung, "Wafer-scale fabrication of single polypyrrole nanoribbon-based ammonia sensor," *Journal of Physical Chemistry C,* vol. 114, no. 25, pp. 11103–11108, 2010.
12. Y. Wang, and G. Li, "Simulation of a silicon nanowire FET biosensor for detecting biotin/streptavidin binding," in *Proceedings of the IEEE 10th International Conference on Nanotechnology*, Seoul, Korea, 2010.
13. Y. Wang, and G. Li, "Performance investigation for a silicon nanowire FET biosensor using numerical simulation," in *IEEE Nanotechnology Materials and Devices Conference*, Monterey, California, 2010.
14. N. A. Lapin, and Y. J. Chabal, "Infrared characterization of biotinylated silicon oxide surfaces, surface stability, and specific attachment of streptavidin," *Journal of Physical Chemistry B,* vol. 113, no. 25, pp. 8776–8783, 2009.
15. S. M. Sze, *Physics of Semiconductor Devices,* New York: Wiley, 1981.
16. D. A. McQuarrie, *Statistical Mechanics,* Sausalito: University Science Books, 2000.
17. K. E. Forsten, R. E. Kozack, D. A. Lauffenburger, and S. Subramaniam, "Numerical solution of the nonlinear Poisson-Boltzmann equation for a membrane-electrolyte system," *Journal of Physical Chemistry,* vol. 98, no. 21, pp. 5580–5586, 1994.
18. D. J. Winzor, "Protein charge determination," *Current Protocols in Protein Science,* 2001, DOI: 10.1002/0471140864.ps0210s41.
19. H.-D. Jakubke, and H. Jeschkeit, *Amino Acids, Peptides and Proteins: An Introduction,* New York: Wiley, 1977.
20. R. Barbucci, A. Magnani, C. Roncolini, and S. Silvestri, "Antigen–antibody recognition by Fourier transform IR spectroscopy/attenuated total reflection studies: Biotin–avidin complex as an example," *Biopolymers,* vol. 31, no. 7, pp. 827–834, 1991.
21. J. Schiewe, Y. Mrestani, and R. Neubert, "Application and optimization of capillary zone electrophoresis in vitamin analysis," *Journal of Chromatography A,* vol. 717, no. 1–2, pp. 255–259, 1995.
22. B. Skoog, and A. Wichman, "Calculation of the isoelectric points of polypeptides from the amino acid composition," *TrAC Trends in Analytical Chemistry,* vol. 5, no. 4, pp. 82–83, 1986.

23. J. C. Gordon, J. B. Myers, T. Folta, V. Shoja, L. S. Heath, and A. Onufriev, "H++: A server for estimating pKas and adding missing hydrogens to macromolecules," *Nucleic Acids Research,* vol. 33, Suppl. 2, pp. W368–W371, 2005.

24. http://biophysics.cs.vt.edu/H++.

25. http://www.nextnano.de.

26. S. Birner, C. Uhl, M. Bayer, and P. Vogl, "Theoretical model for the detection of charged proteins with a silicon-on-insulator sensor," *Journal of Physics: Conference Series,* vol. 107, no. 1, p. 012002, 2008.

27. http://nanohub.org/resources/NANOFINFET.

3

Modeling and Simulation of Organic Photovoltaic Cells

Guangyong Li, Liming Liu, and Fanan Wei

CONTENTS

3.1 Introduction

Organic photovoltaic cells have great potential to enable mass production with extremely low cost. The power conversion efficiency of organic photovoltaic cells has improved rapidly in the past few years, exceeding 5% in 2005 [1,2] and 9.8% in 2012 [3]. But it is still low compared to other traditional solar cells. The relatively low efficiency is rooted in the low dielectric constants of organic semiconductors, in which tightly bonded electron–hole pairs (excitons) instead of free electron–hole pairs are formed after light absorption [4]. The exciton can be efficiently separated only into a free electron and hole at the interface between the donor and acceptor materials by the built-in potential, arising from the offset of Fermi levels between donor and acceptor materials. Because the exciton diffusion length of organic semiconductors is very short (about 10 nm [5]), only excitons created within 10 nm from the donor–acceptor (D–A) interface can be dissociated; all others decay to ground states.

The major milestone of organic photovoltaic cells was the introduction of bulk heterojunction (BHJ) configuration (Figure 3.1) reported in 1995 [6]. In BHJ configuration, the donor and acceptor materials are blended together to form phase-separated nanostructures (nanoscale morphology). The nanoscale morphology of the disordered bulk heterojunctions plays a critical

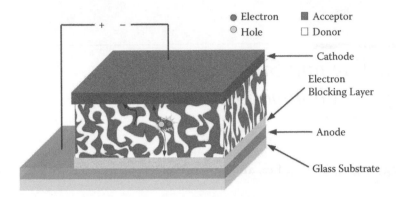

FIGURE 3.1
Configuration of organic photovoltaic cells made from BHJs.

role in determining the overall device performance. The nanoscale morphology of BHJs must be controlled in such a balanced way that the formation of phase-separated domains can provide enough interconnections for transporting free carriers but does not sacrifice too much interfacial contact areas for efficient exciton dissociation; the thickness of BHJs must be chosen such that it is thin enough for efficient collection of free carriers but it is thick enough to absorb sufficient light. Even though the world-record efficiency of organic photovoltaic cells has been close to 10% [3], only 3%–5% is routinely achievable in general labs and less than 1% in mass production because of the lack of a valid tool to optimize the device design and the fabrication process. Compared to trial-and-error-based physical experiments, computational modeling and simulation provide more effective and economical alternatives to optimize the device for best performance. This chapter will provide the technical details on how to model and simulate the operation of organic photovoltaic cells.

3.2 Fundamentals of Organic Photovoltaic Cells

In this section, we briefly introduce the fundamentals of organic photovoltaic cells. More details on device physics can be found in the review article by Deibel et al. [7].

After light absorption, tightly bound electron-hole pairs (excitons) are created in organic semiconductors because of their low dielectric constants. In order to convert the solar energy into electricity, excitons must be separated into free carriers (electrons and holes) before they decay to ground states. The most efficient way for exciton dissociation is through a BHJ configuration (Figure 3.1) in which the electron donor and acceptor materials are blended into an intimately mixed nanostructure with the characteristic phase separation length scale of the order of 10–20 nm. Because the size of the phase-separated domains in the blend is similar to the exciton diffusion length, exciton decay can be significantly reduced because of the sufficient interfaces in the proximity of every generated exciton. For a polymer BHJ configuration with proper nanoscale morphology, the exciton dissociation is so sufficient that the exciton dissociation efficiency is close to 100% [8]. The disadvantage of a BHJ configuration is that it requires percolated pathways for the charge carriers transporting to their corresponding electrodes. Even though there are other kinds of configurations for organic photovoltaic cells, such as bilayer, we mainly discuss the BHJ configuration because of its obviously better performance. By default, in this chapter organic photovoltaic cells are referred to as those with a BHJ configuration.

The complete operation process of organic photovoltaic cells can be summarized as follows: the incident light is absorbed in the active layer, and tightly bounded excitons (electron–hole pair) are generated (Figure 3.2a); the excitons then diffuse to the D–A interface and dissociate into charge-transfer excitons (polarons) [9], which can be further separated into free electrons and holes (Figure 3.2b). Subsequently, these free charge carriers are then

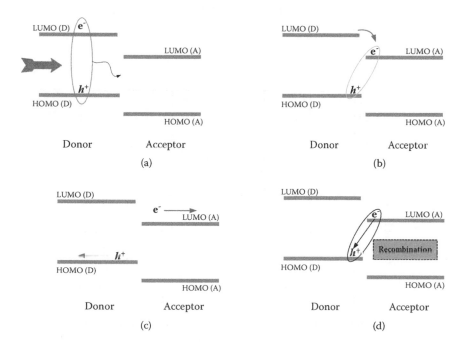

FIGURE 3.2
Schematic diagram to show the basic working principle of organic photovoltaic cells: (a) exciton formation and diffusion; (b) exciton dissociation; (c) free carrier transport; (d) geminate recombination.

transported to and finally collected by the corresponding electrodes, respectively (Figure 3.2c); During the exciton diffusion and charge carrier transport processes, recombination losses will occur. The excitons may decay to ground states before they reach the D–A interface (not shown); the charge transfer states may also undergo geminate recombination (Figure 3.2d). The charge carriers could be further lost during transport through either monomolecular or bimolecular recombination (not shown in the figure).

The power conversion efficiency of any photovoltaic cells can be obtained from their current–voltage (J–V) characteristics under standard illumination. The typical J–V curves of a photovoltaic device under illumination (solid line) are simulated as shown in Figure 3.3. The open circuit voltage V_{oc} is the maximum voltage that can be generated in the cell, and corresponds to the voltage where current under illumination is zero. The current density that can run through the cell at zero applied voltage is called the *short circuit current* J_{SC}. The maximum electrical power P_{max} of a photovoltaic cell is located in the fourth quadrant where the product of current density J and voltage V reaches its maximum value. The ratio between P_{max} and $V_{oc} \times J_{SC}$ represents the quality of the J–V curve shape and is defined as the fill factor (*FF*)

$$FF = \frac{V_{max} \times J_{max}}{V_{oc} \times J_{sc}} \tag{3.1}$$

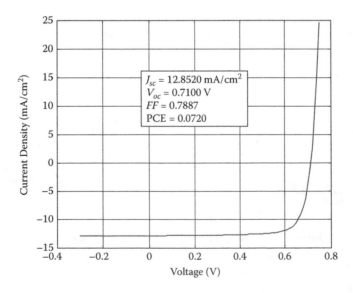

FIGURE 3.3
J–V characteristic curves of the typical PV device under illumination.

The power conversion efficiency is a common parameter to characterize the overall performance of a solar cell, but it is not convenient to analyze the device performance in detail, especially for organic photovoltaic cells with BHJ. External quantum efficiency (η_{EQE}), defined as the ratio of the number of charge carriers collected by the solar cell to the number of photons at a given energy shining on the solar cell from outside (incident photons), is another common parameter to characterize the performance of a solar cell responding to light at a certain wavelength, which has been frequently used to analyze the device performance. The external quantum efficiency of organic photovoltaic cells can be expressed by the product of the efficiencies of four sequential steps through the following formula [10]:

$$\eta_{EQE} = \eta_A \eta_{ED} \eta_{DA} \eta_{CT} \eta_{CC} \tag{3.2}$$

where η_A represents the efficiency of photon absorption leading to the generation of excitons; η_{ED} represents the efficiency of exciton diffusion to the D–A interface; η_{DA} represents the efficiency of exciton dissociation at the D–A interface; η_{CT} represents the efficiency of free charge transport to the electrodes; and η_{CC} represents the efficiency of the charge collection at the electrodes. Frequently, we use the following formula to calculate external quantum efficiency of any solar cells for simplicity:

$$\eta_{EQE} = \eta_A \eta_G \eta_C \tag{3.3}$$

where η_G is the free carrier generation efficiency with $\eta_G = \eta_{ED} \cdot \eta_{DA}$ and η_C is the free carrier collection efficiency with $\eta_C = \eta_{CT} \cdot \eta_{CC}$. Another important parameter to characterize the performance of solar cells is the internal quantum efficiency (η_{IQE}), defined as the ratio of the number of charge carriers collected by the solar cell to the number of photons absorbed by the active layer, which can be calculated by $\eta_{IQE} = \eta_G \cdot \eta_C$.

3.3 Optical Modeling

The purpose of optical modeling is to calculate the light absorption and exciton creation based on materials properties and the structure of organic photovoltaic cells. In this section, we first introduce the solar spectrum, and then describe the calculation of light absorption for homogeneous layers by the optical transfer matrix theory.

The surface temperature of the sun is around 5800 K, so the radiation spectrum from the sun is similar to the blackbody radiation at 5800 K. The

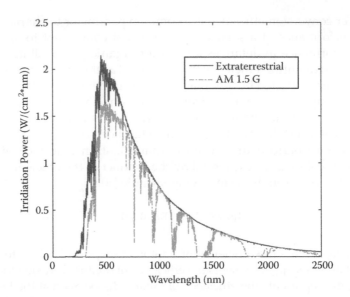

FIGURE 3.4
Solar spectrum. The solid curve is extraterrestrial measurement data and the dotted curve represents AM 1.5G spectrum. (Original data from http://rredc.nrel.gov/solar/spectra/.)

radiation power density of the sun on the external atmosphere of the Earth (without considering absorption of the Earth's atmosphere) is

$$S(\lambda, 5800) = \frac{2\pi hc^2}{\lambda^5} \cdot \frac{1}{\exp(hc/\lambda k * 5800K) - 1} \cdot \left(\frac{r_s}{d_{SE}}\right)^2, \tag{3.4}$$

where r_s is radius of the sun, $r_s = 6.96*10^8$ m; and d_{SE} is the distance between the sun and the Earth, $d_{SE} = 1.496*10^{11}$ m. For the characterization of ordinary solar cells on the Earth, the incident spectrum AM 1.5 global is used because the Earth's atmosphere absorbs incident light with a certain wavelength, which results in the deviation of the solar spectrum on the Earth from 5800 K blackbody radiation. In the following discussion, we use the AM 1.5 global spectrum (dotted curve; certified by the National Renewable Energy Laboratory), in which there are some distinct features corresponding to the specific absorption of the Earth's atmosphere.

Traditionally, the Beer–Lambert law is often used to describe the light intensity in bulk materials, assuming an exponential decay as

$$I(x) = I_0 exp(-\alpha x) \tag{3.5}$$

where $I(x)$ is the light intensity at position x, I_0 is the incident light intensity, and α is the absorption coefficient. However, light absorption in organic

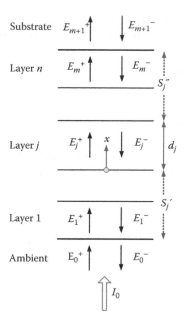

FIGURE 3.5
Schematic of *m*-layer structure between ambient and substrate.

photovoltaic cells is greatly affected by optical interference given that the thickness of each layer is less than the light wavelength. Figure 3.5 shows optical transport in a typical multilayer solar cell. A plane wave is incident from the left at an *m*-layer structure (layer 1 to layer *m*) between a semi-infinite transparent ambient and a semi-infinite opaque substrate. Due to the interference effect in the thin-film device, the optical electric field at any point x in layer j is a complex quantity and consists of a positive component $E_j^+(x)$ and a negative component $E_j^-(x)$. The optical properties of each layer (j) are described by the thickness d_j and the complex index of refraction as

$$\tilde{n}_j = n_j + i\kappa_j. \tag{3.6}$$

Since the thickness of thin films in organic photovoltaic cells is generally less than the wavelength of incident light, the optical effects such as reflections and interference are important and cannot be neglected when evaluating the electromagnetic field inside the device. The behavior of the light in the devices or the electromagnetic field distribution can be calculated either from an optical transfer matrix or from Maxwell equations. Optical transfer matrix theory is an approach to model the light propagation through a one-dimensional layered stack of different materials. To apply optical transfer matrix theory in the simulation of optical absorption and light intensity distribution within organic thin film solar cell, several assumptions need to be made [11]:

1. Layers included in the device are considered to be homogeneous and isotropic, so that their linear optical response can be described by a scalar complex index of refraction.
2. Interfaces are parallel and flat compared to the wavelength of the light.
3. The incident light at the device can be described by plane waves.

For most organic photovoltaic cells without texturing on each layer, these conditions hold. Based on optical transfer matrix theory, the light behavior at the interface between j and $k = j + 1$ layers can be described by a 2×2 interface matrix containing the complex Fresnel coefficients. The interface matrix I_{jk} can be expressed as

$$I_{jk} = \begin{bmatrix} (\tilde{n}_j + \tilde{n}_k)/2\tilde{n}_j & (\tilde{n}_j - \tilde{n}_k)/2\tilde{n}_j \\ (\tilde{n}_j - \tilde{n}_k)/2\tilde{n}_j & (\tilde{n}_j + \tilde{n}_k)/2\tilde{n}_j \end{bmatrix}. \tag{3.7}$$

The light propagating in layer j can be described by the layer matrix L_j as

$$L_j = \begin{bmatrix} \exp\left(-i\dfrac{2\pi\tilde{n}_j}{\lambda} \cdot d_j\right) & 0 \\ 0 & \exp\left(-i\dfrac{2\pi\tilde{n}_j}{\lambda} \cdot d_j\right) \end{bmatrix} = \begin{bmatrix} \exp(-ik_j \cdot d_j) & 0 \\ 0 & \exp(-ik_j \cdot d_j) \end{bmatrix}, \tag{3.8}$$

where $k_j = 2\pi\tilde{n}_j/\lambda$. From L_j in Equation 3.8, the real part of complex refraction, n_j, responses for the phase propagation and image part, κ_j, are related to the light absorption. The optical electric field in the ambient (subscript 0) is related to that in the substrate (subscript $m + 1$) by the total transfer matrix S,

$$\begin{bmatrix} E_0^+ \\ E_0^- \end{bmatrix} = S \begin{bmatrix} E_{m+1}^+ \\ E_{m+1}^- \end{bmatrix}, \tag{3.9}$$

where the total optical transfer matrix S is the product of all interface and layer matrices expressed as

$$S = \begin{bmatrix} S_{11} & S_{12} \\ S_{21} & S_{22} \end{bmatrix} = \left(\prod_{v=1}^{m} I_{(v-1)v} L_v\right) I_{m(m+1)}. \tag{3.10}$$

The optical electric field at a distance x within layer j can be expressed as

$$E_j(x) = E_j^+(x) + E_j^-(x). \tag{3.11}$$

To express $E_j(x)$ in terms of known quantities, Equation 3.11 is split into two partial transfer matrices S_j' and S_j''. As illustrated in Figure 3.5, $S = S_j' L_j S_j''$. These two matrices can be expressed as

$$S_j' = \begin{bmatrix} S_{j11}' & S_{j12}' \\ S_{j21}' & S_{j22}' \end{bmatrix} = \left(\prod_{v=1}^{j-1} I_{(v-1)v} L_v \right) I_{(j-1)j}, \tag{3.12}$$

$$S_j'' = \begin{bmatrix} S_{j11}'' & S_{j12}'' \\ S_{j21}'' & S_{j22}'' \end{bmatrix} = \left(\prod_{v=j+1}^{m} I_{(v-1)v} L_v \right) I_{m(m+1)}. \tag{3.13}$$

For the partial system transfer matrices for layer j, we have:

$$\begin{bmatrix} E_0^+ \\ E_0^- \end{bmatrix} = S_j' \begin{bmatrix} E_j'^+ \\ E_j'^- \end{bmatrix}, \tag{3.14}$$

$$\begin{bmatrix} E_j''^+ \\ E_j''^- \end{bmatrix} = S_j'' \begin{bmatrix} E_{m+1}^+ \\ E_{m+1}^- \end{bmatrix}, \tag{3.15}$$

where $E_j'^+$ and $E_j'^-$ refer to the left boundary $(j-1)j$ in layer j and $E_j''^+$ and E_j' refer to the right boundary $j(j+1)$ in layer j. After algebraic manipulation, Equation 3.11 can be expressed as

$$E_j(x) = E_j^+(x) + E_j^-(x) = t_j^+ \left[e^{ik_j x} + r_j'' e^{ik_j(2d_j - x)} \right] E_0^+, \tag{3.16}$$

where E_0^+ is the electric field of the incident light from the ambient, where

$$t_j^+ = \left[S_{j11}' + S_{j12}' r_j'' e^{2ik_j d_j} \right]^{-1}, \tag{3.17}$$

$$r_j'' = S_{j21}''/S_{j11}''. \tag{3.18}$$

For a detailed derivation, please refer to [11]. After determining the distribution of optical electric field $E_j(x)$ in layer j as Equation 3.16, the light intensity at a distance x within layer j of the device can be expressed as

$$I_j(x, \lambda) =$$

$$T_j I_0(\lambda) \left[e^{-\alpha_j x} + \rho_j''^2 \cdot e^{-\alpha_j(2d_j - x)} + 2\rho_j'' \cdot e^{-\alpha_j d_j} \cdot \cos\left(\frac{4\pi n_j}{\lambda}(d_j - x) + \delta_j'' \right) \right], \tag{3.19}$$

where $I_0(\lambda)$ is the intensity of the incident light, $T_j = (n_j/n_0)t_j^2$ is the internal intensity transmittance, and ρ_j'' and δ_j'' are the absolute value and the argument of the complex reflection coefficient. The first term inside the brackets of Equation 3.19 on the right-hand side originates from the optical electric field propagating in the positive x direction $E_j^+(x)$, the second term from the field propagating in the negative x direction $E_j^-(x)$, and the third term arises from the optical interference of the first two terms. The third term becomes especially important when the thickness of thin film is comparable with the light wavelength. When the thickness of an active layer is much larger than the light wavelength ($d_j \gg \lambda$), the third term is negligible, and only the first two terms dominate, converging to the Beer–Lambert law for bulk materials.

Once the distribution of light intensity in the active layer is determined, the rate of energy dissipated per unit volume Q can be determined as

$$Q(x,\lambda) = \alpha(\lambda)I(x,\lambda) \tag{3.20}$$

where $\alpha(\lambda) = 2\pi\kappa/\lambda$ is the absorption coefficient of the active layer. Then the density of photon absorbed in the active layer is

$$n(x) = \int \frac{\lambda}{hc}Q(x,\lambda)d\lambda, \tag{3.21}$$

where h is Planck's constant, and c is light speed in a vacuum.

The photon absorption versus the thickness of the active layer for the AM 1.5G spectrum is shown in Figure 3.6. It can be seen that an optical interference effect exists if the optical transfer matrix method (solid curve) is applied and as a result, the absorption is enhanced at a certain thickness. The calculation results from the Beer–Lambert law (dashed curve); however, it shows no sign of oscillation. The simulation results based on the optical transfer matrix indicate that the optimal thickness of the active layer is approximately either 90 nm or 220 nm, coincident with experimental observations [12,13].

3.4 Electrical Modeling and Simulation by Drift-Diffusion Model

The full current–voltage (J–V) characteristics of organic photovoltaic cells can be simulated by an equivalent-circuit model. Considering that such a model hardly reveals any fundamental physics of the devices, it is not discussed. Instead, we review how to simulate full current–voltage (J–V) characteristics of organic photovoltaic cells by the continuum drift-diffusion

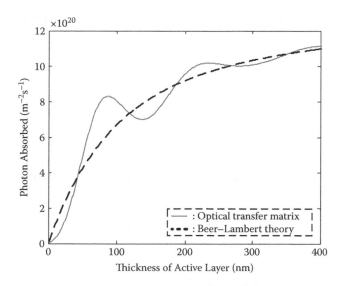

FIGURE 3.6
Calculated photons absorbed of AM 1.5G spectrum in the P3HT/PCBM solar cell versus its thickness.

model, which reveals more physics of the organic photovoltaic cells. The drift-diffusion model is derived from Poisson's equation and continuity equations for electrons and holes, which are commonly used for describing the electrical behavior of semiconductor devices:

$$\frac{\partial^2}{\partial x^2}\phi(x) = \frac{q}{\varepsilon}\left[n(x) - p(x)\right]$$

(3.22)

$$\frac{\partial n}{\partial t} = G - R_n + \frac{1}{q}\frac{\partial J_n(x)}{\partial x}$$

(3.23)

$$\frac{\partial p}{\partial t} = G - R_p - \frac{1}{q}\frac{\partial J_p(x)}{\partial x},$$

(3.24)

where ϕ represents the electric potential; q is the unit charge; ε is the dielectric constant of the semiconductor; n and p are the electron and hole density, respectively; $J_{n,\,p}$ is the electron (hole) current; G is the generation rate; and $R_{n(p)}$ is the recombination rate. Equation 3.22 is the Poisson equation and Equations 3.23 and 3.24 are the continuity equations. The full J–V characteristics of organic photovoltaic cells can be obtained by solving these equations when the mean effects of the band gap and bulk mobility are considered.

For steady-state analysis ($\partial n/\partial t = \partial p/\partial t = 0$), Equations 3.23 and 3.24 can be simplified as

$$\frac{\partial J_n(x)}{\partial x} = q(-G + R_n) \tag{3.25}$$

$$\frac{\partial J_p(x)}{\partial x} = q(G - R_p). \tag{3.26}$$

J_n (J_p) can be written as functions of ϕ, n and p, consisting of drift and diffusion components:

$$J_n = -qn(x)\mu_n \frac{\partial}{\partial x}\phi(x) + qD_n \frac{\partial}{\partial x}n(x) \tag{3.27}$$

$$J_p = -qp(x)\mu_p \frac{\partial}{\partial x}\phi(x) - qD_p \frac{\partial}{\partial x}p(x), \tag{3.28}$$

where $\mu_{n(p)}$ is the electron (hole) mobility and $D_{n(p)}$ is the electron (hole) diffusion coefficient. By taking Equations 3.27 and 3.28 back to Equations 3.25 and 3.26, we can have the two following carrier continuity equations:

$$-n(x)\mu_n \frac{\partial^2\phi}{\partial x^2} - \frac{\partial n}{\partial x}\mu_n \frac{\partial\phi}{\partial x} + D_n \frac{\partial^2 n}{\partial x^2} = -G + R_n \tag{3.29}$$

$$-p(x)\mu_p \frac{\partial^2\phi}{\partial x^2} - \frac{\partial p}{\partial x}\mu_p \frac{\partial\phi}{\partial x} - D_p \frac{\partial^2 p}{\partial x^2} = G - R_p \tag{3.30}$$

The goal of the simulation is to calculate carrier density $n(x)$, $p(x)$ as well as electric potential $\phi(x)$ and then obtain the current with respect to the externally applied voltage (J–V characteristics). To simulate organic photovoltaic cells using the above model, several critical issues need to be clarified considering the unique properties of BHJ.

Permittivity: The permittivity ε is a spatially averaged dielectric constant of donor/acceptor blend, and hence ε is different for blends with different weight ratios between donor and acceptor materials [14].

Mobility: The bulk carrier (or apparent) mobility (μ_e for electrons and μ_h for holes) in the BHJ configuration, dependent on the nanoscale morphology in the disordered blend, is different from the value in its pure phase [15]. In addition to the dependence on nanoscale morphology and disorder, the apparent mobility is also largely dependent on the electrical field and free carrier concentration [16]. The apparent mobility further affects other aspects of the device such as recombination, short-circuit current, and open-circuit voltage, and thus the device performance [17]. Fortunately, the dependence on electrical field and carrier concentration is considered as weak at low electrical fields (less than 10^7 V/m) and low carrier concentration [16]. Therefore, it is usually safe to assume a constant mobility from short-circuit to open-circuit ranges.

Open Circuit Voltage: Open-circuit voltage is limited by the polaron pair energy [18,19] and also weakly affected by the work functions of the electrodes [20]. Therefore, it is not difficult to estimate the exact open-circuit voltage. In the simulation, the common approach to estimate the Voc is taking an offset from the effective bandgap, expressed as $Voc = LUMO_a - HOMO_d - \Delta$ [21].

Einstein Relation: Traditionally, the classic Einstein relation ($D_{n,p}/\mu_{n,p} = kT/q$) is assumed. However, deviation from the Einstein relation in disordered semiconductors has been reported [22–24]. A generalized Einstein relation can be used for disordered organic materials with modifications that the LUMO energy level is from the acceptor and the HOMO energy level is from the donor [25].

Boundary Conditions: Dirichlet boundary conditions [12,21] with ohmic contacts are assumed, where the surface potential and carrier concentrations are fixed at both ends.

Recombination: The determination of dominant loss mechanism in organic photovoltaic cells is very critical to correctly model and simulate their operation. Unfortunately, the exact mechanism of the recombination is not clear yet. It is commonly accepted that the recombination in organic photovoltaic cells is not a Langevin-type bimolecular recombination, but its true mechanism is still under heated debate. The majority believe bimolecular recombination is dominant, even though it has been suppressed compared to a Langevin-type recombination. The reduced bimolecular recombination is supported not only by direct experimental studies [26–29] but also by deviation of the ideality factor [30] and by fitting simulation to light-intensity dependent *J–V* characteristics [31]. Recently, several researchers, including the authors, have suggested that monomolecular recombination may dominate in polymer BHJ solar cells [32–34]. This argument is supported by studying the light-intensity dependent *J–V* characteristics for regions below the open-circuit condition. To reconcile the two contradictory opinions (either monomolecular or bimolecular), Cowan et al. proposed that the recombination loss is voltage dependent and evolves from monomolecular recombination at the short-circuit condition to bimolecular recombination at the open-circuit condition [35], which sounds more reasonable.

TABLE 3.1

Parameter Normalization

Parameter	Normalization Parameter	Unit	After Normalization
Position (x)	$L_D = (\epsilon kT/q^2 n_i)^{1/2}$	cm	$X = X/L_D$
Potential (φ)	$V_{kT} = kT/q$	V	$\varphi = \varphi/V_{kT}$
Carrier density (n, p)	n_i	cm^{-3}	$n = n/n_i$
Diffusion $(D_{n,p})$	D_0	cm^2/s	$D_{n,p} = D_{n,p}/D_0$
Mobility $(\mu_{n,p})$	$\mu_0 = qD_0/kT$	cm^2/(Vs)	$\mu_{n,p} = \mu_{n,p}/\mu_0$
Time (t)	$t_0 = L_D^2/D_0$	s	$t = t/t_0$
Recombination rate (R)	$R_0 = D_0 n_i/L_D^2$	1/(cm^{3}s)	$R = R/R_0$
Current density $(J_{n,p})$	$J_0 = qD_0 n_i/L_D$	A/cm^2	$J = J/J_0$

Source: L. Liu, and G. Li, "Modeling and simulation of organic solar cells," in *IEEE Nanotechnology Materials and Devices Conference*, Monterey, California, 2010.

Because of the coupling between the electric potential φ and carrier density $n(p)$ in the drift current, it is impractical to solve differential equations (3.22–3.26) analytically. Therefore, a numerical method must be used to solve coupled differential equations. The first step is parameter normalization, which simplifies the calculation in the numerical simulation, especially for Poisson and continuity equations in which there are some constant parameters. The normalized parameters include Debye length, intrinsic carrier density (n_i), thermal energy (kT/q), and others, as shown in Table 3.1.

After normalization, the second step is to discretize the electric potential φ and carrier density $n(p)$ on a simulation grid so that the continuous functions can be represented by vectors of function values at the nodes, and the differential operator is replaced by suitable difference operators. A one-dimensional ununiformed grid is shown in Figure 3.7. The region considered here ($x = 0$ to $x = W$) has been discretized to N_w nodes, which were numbered as 1, 2, ..., N, ..., N_w. The middle point between the N node and $N + 1$ node was named M. Therefore, there are totally $N_w - 1$ middle points between the nearest nodes, named 1, 2, ..., M, ..., M_w ($M_w = N_w - 1$). The distance between

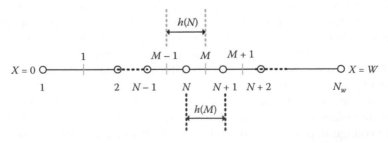

FIGURE 3.7
Schematic of one-dimensional discretization.

the N node and $N + 1$ node is $h(M)$, and the distance between the middle point $M - 1$ and M is $h(N)$. The discretization of electric potential φ using the central differential formula can be expressed as

$$\partial\varphi(N)/\partial x = [\varphi(M)-\varphi(M-1)]/h(N) \tag{3.31}$$

The discretization of carrier density $n(p)$ requires more care because the change of electric potential φ in the semiconductor is gradual, but the change of carrier density $n(p)$ is rapid, and hence the general calculation method will always result in a nonconverge situation. Therefore, the Scharfetter–Gummel method was applied to discretize the carrier density n as [37]

$$n(x \in [N, N+1]) = [1 - g_i(x)] * n(N) + g_i(x) * n(N+1) \tag{3.32}$$

$$p(x \in [N, N+1]) = [1 - g_i(x)] * p(N) + g_i(x) * p(N+1), \tag{3.33}$$

where the pre-factor $g_i(x)$ is

$$g_i(x) = \left[1 - \exp\left(\frac{\phi(N+1) - \phi(N)}{kT} * \frac{x - x(N)}{h(N)}\right)\right] \bigg/ \left[1 - \exp\left(\frac{x - x(N)}{h(N)}\right)\right]. \tag{3.34}$$

It is clear that in the middle between the nearest grid points, the carrier density has a nonlinear relation with distance x and the continuity equations are coupled together. In addition, the change of electric potential in the semiconductor is gradual, but the change of carrier density is very rapid, and hence the general calculation method will not always converge. After discretization of φ and n (p) by the central difference formula and by the Scharfetter and Gummel method [37], respectively, either a decoupled iteration method [38] (Gummel's method) or a coupled method [39] (Newton method) can be employed to solve the linearized and discretized differential equations. A detailed description regarding discretization and linearization, as well as Gummel's method and the Newton method, can be found in Reference 40.

With these simulation algorithms, we can develop a simulation tool that is able to simulate the complete $(J–V)$ characteristics of organic photovoltaic cells by considering the optical absorption under standard illumination as shown in Figure 3.3. With this simulation, we can also study the device physics of organic photovoltaic cells.

3.4 Electrical Modeling and Simulation by Monte Carlo Model

The nanoscale morphology of disordered bulk heterojunctions plays a critical role in determining overall device performance. It has been found that carrier mobility can be improved by ameliorating the nanoscale morphology of the active layer through thermal annealing [1,41–43] and solvent annealing [2,43,44] to improve the crystallinity and phase separation, which have been proven with XRD (x-ray diffraction) and TEM (transmission electron microscopy) [1]. Unfortunately, macroscopic simulation using the drift-diffusion model to derive Poisson's equation and continuity equations [5–8] only considers the mean effect of many factors without correlating the device performance to the nanomorphology. Mesoscopic approaches, such as Monte Carlo simulation, on the other hand, can help gain a deeper understanding of the relationship between the nanoscale morphology of the active layer and the performance of the device.

In order to study the relationship between the morphology and the performance of organic photovoltaic cells, a series of blend morphologies with different phase separations needs to be generated. In a P3HT:PCBM system with a weight ratio of 1:1, the scales of phase separation are characterized by domain size $a = 3V/A$, where V is the total active layer volume, and A is donor/acceptor interfacial area. Herein, the Ising model [45] is often adopted to obtain the nanoscale morphology. The morphology generation process is shown as follows:

First, the active layer is discretized into many small cubes (a lattice constant of 1 nm is used here). Every lattice, also called a *site*, is assigned a value 0 or 1 with equal probability, where 0 stands for donor and 1 for acceptor.

Second, choose two neighboring sites randomly from all the lattices and calculate the Ising Hamiltonian value of the system comprised of the chosen two sites, their neighbors, and second-nearest neighbors according to the following equation:

$$\varepsilon_i = -\frac{J}{2}\sum_j (\delta_{s_i,s_j} - 1), \tag{3.35}$$

where δ_{s_i,s_j} is the delta function and $J = +1.0 \, K_b T$. Here, K_b represents the Boltzmann constant and T is the temperature. The contribution of the second-nearest neighbors is scaled by a factor of $1/\sqrt{2}$.

Third, exchange the spin value (0 or 1) of the chosen two sites with the probability

$$P(\Delta\varepsilon) = \frac{\exp(-\Delta\varepsilon/K_B T)}{1+\exp(-\Delta\varepsilon/K_B T)}, \tag{3.36}$$

where $\Delta\varepsilon$ is the difference between Hamiltonian values before and after exchange.

Finally, after a great number of attempted spin exchanges, a morphology series with the required domain sizes can be generated as shown in Figure 3.8a and 3.8b.

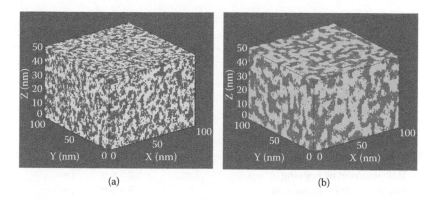

FIGURE 3.8
Morphologies generated by the Ising model for P3HT:PCBM devices: nanoscale morphology of BHJs with weight ratio of 1:1 and with different phase separations at (a) 5 nm and (b) 10 nm.

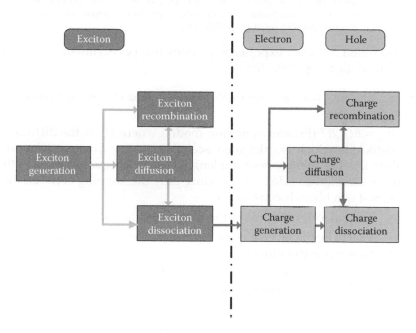

FIGURE 3.9
The major events during the operation process of organic photovoltaic solar cells.

Three kinds of particles (exciton, electron, and hole) are included in the model of polymer BHJ solar cells. Careful analysis of the mechanism of photocurrent generation will result in ten types of events as shown in Figure 3.9.

The most important parameters to be set are the attempt-to-happen rates of all the events considered. The values of these constants for the P3HT:PCBM system used in the simulation are listed in Table 3.2. They are

TABLE 3.2

Event Happen Rate Constants

Events	Happen Rate Constants	Value	Unit
Absorbing photon to generate exciton	w_{exg}	4×10^{-12}	$ps^{-1}nm^{-3}$
Exciton diffusion via hopping	w_{exh}	2×10^{-1}	ps^{-1}
Exciton decay by recombination	w_{exd}	2×10^{-3}	ps^{-1}
Exciton separate at the D/A interface	w_{exs}	20.00	ps^{-1}
Electron transport by hopping	w_{elh}	1.682	ps^{-1}
Hole transport by hopping	w_{hoh}	2.5×10^{-1}	ps^{-1}
Electron/hole recombination	w_{ehr}	5×10^{-7}	ps^{-1}
Electron extracted by electrode	w_{elc}	1×10^{-2}	ps^{-1}
Hole extracted by electrode	w_{hoc}	1×10^{-2}	ps^{-1}

Source: Data from M. Casalegno, G. Raos, and R. Po, "Methodological assessment of kinetic Monte Carlo simulations of organic photovoltaic devices: The treatment of electrostatic interactions," *Journal of Chemical Physics*, vol. 132, no. 9, p. 094705, 2010.

obtained based on either experienced values from experiments or on theoretical calculations, specifically:

- w_{exg}, w_{ehr}, w_{elc} and w_{hoc} are experienced values chosen according to experiments.
- $W_{exh} = 6D_{ex}/L^2$ (Brownian motion model) where D_{ex} is the diffusion coefficient, which is calculated according to the Einstein relationship; L is the average hopping length, which is equal to the length of smallest lattice side $l = 1$ nm, since only the hopping between the nearest neighbors is considered.
- w_{exd} is equal to $1/\tau$, where τ is the lifetime of the exciton.
- w_{exs} is set to be the inverse of the exciton dissociation time at the D/A interface, while 0 at other sites.
- $w_{elh(hoh)}$ can be calculated by combining the Brownian motion model and the Einstein relationship as

$$w_{elh(hoh)} = \frac{6k_B T\mu_{e^-(h^+)}}{e\ell^2}\exp(2\gamma\ell),$$

where μ is the mobility, e is the unit charge, and γ is the localization constant valued at 2.0 nm^{-1}.

- The relative permittivity of the active layer is set to be a spatial mean value.

Figure 3.10a shows the simulation results that excitons were generated at random sites, then the dissociation of excitons, and the hopping of the generated

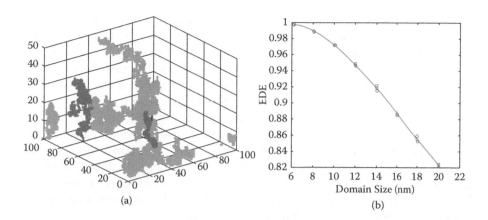

(a) (b)

FIGURE 3.10
Simulation of exciton and charge carriers in BHJ blends with phase-separation size 5 nm (a) and simulation of exciton dissociation efficiency for BHJ blend with phase-separation size 5 nm (b).

electrons (dark) and holes (gray). The influence of morphology in the blend on exciton dissociation efficiency is shown in Figure 3.10b. It can be seen that smaller domain size yields higher exciton dissociation efficiency.

Monte Carlo models can effectively account for the phase-separated morphology [47], but they mainly focus on one or a few aspects of the device. Even though Monte Carlo is capable of simulating a few points of the *J–V* characteristics [48], it is not convenient to simulate the full experimental *J–V* characteristics. Even though the results from Monte Carlo simulation now meet our expectations from our understanding of polymer BHJ, their validity and accuracy need to be further investigated considering the following limitations: (1) the morphology generated by the Ising model hardly resembles the real morphology of a device, which is much more complex; (2) there is a lack of effective tools to characterize the nanoscale morphology and thus to validate the results from Monte Carlo simulation through experiments; and (3) we have limited understanding of the working principles of polymer BHJ. Because Monte Carlo simulation is fundamentally a statistical approach, its accuracy also is dependent on the sample size used in the calculation. A larger sample size will increase the accuracy but may require unaffordable computational time. Therefore, the validity and accuracy need to be much improved in future research.

3.5 Discussion and Conclusion

In this chapter, we first provide background on the fundamentals of organic photovoltaic cells from basic configurations and operations to device performance, as well as the nanoscale morphology that dictates the performance of the device. We then introduce optical modeling for calculation of light absorption based on transfer matrix theory that accounts for the reflection and interference in each layer. We further present electrical modeling for simulating device performance. Two kinds of electrical modeling approaches are discussed: the macroscopic approach using the drift-diffusion model and the mesoscopic approach using a Monte Carlo model. The macroscopic approach based on the drift-diffusion model can predict the overall performance of organic photovoltaic cells by taking into account the mean effects of the devices, but it fails to take nanoscale morphology into account in the model. The mesoscopic approach using the Monte Carlo model does take nanoscale morphology into consideration but it can study only one or a few aspects of organic photovoltaic cells.

We have shown that modeling and simulation of organic photovoltaic cells are very useful in predicting device performance and analyzing working principles. However, the results from modeling and simulation are only meaningful when the modeling and simulation have been proven to be valid and accurate. Valid modeling and simulation have to be built on the correct understanding of the working principles of devices as well as experimental verification. Some of the working mechanisms of organic photovoltaic cells such as recombination are not completely clear; further validation of some modeling and simulation approaches such as Monte Carlo simulation is necessary. The accuracy of modeling and simulation is based on the accuracy of the feed-in parameters. Unfortunately, there is still a lack of effective tools to measure some critical parameters such as nanoscale morphology, energy disorder, and recombination. Therefore, the validation and accuracy of modeling and simulation are the major limiting factors that preclude their wide usage.

To make organic photovoltaic cells commercially viable, there is a pressing need to study modeling and simulation, and thus to optimize organic photovoltaic cells for the best device performance. So far, neither the macroscopic approach based on the drift-diffusion model nor the mesoscopic approach based on Monte Carlo modeling alone is able to exactly describe the full operation of organic BHJ solar cells. The macroscopic approach can simulate the macroscopic performance of the device by considering the mean effect of many factors without directly correlating device performance to nanomorphology. Mesoscopic simulation based on the Monte Carlo model takes nanoscale morphology into consideration to simulate a few important factors such as exciton dissociation, but it fails to simulate the macroscopic performance of the devices because of the requirement for excessive computational power.

Because of the complexity of organic photovoltaic cells, it is not possible to simply tune one or a few parameters to achieve optimal performance. Instead, the optimization of organic photovoltaic cells is a fine balancing act, which means all the parameters need to be tuned simultaneously. Therefore, a multiscale modeling and simulation approach that includes all the aspects of organic photovoltaic cells in the model is needed. For example, Monte Carlo simulation can be used to predict a few properties of BHJs such as exciton dissociation, carrier recombination, and carrier bulk mobility with respect to nanoscale morphology; then these parameters can be fed into a macroscopic simulation based on the drift-diffusion model, which can predict the overall performance of a device. By combining the optical model with these multiscale simulations, the optimal design of the devices such as the thickness of the functional layer and the nanoscale morphology of the BHJs can be identified simultaneously. The optimization of organic photovoltaic cells via multiscale modeling and simulation is discussed in Chapter 4.

References

1. W. L. Ma, C. Y. Yang, X. Gong, K. Lee, and A. J. Heeger, "Thermally stable, efficient polymer solar cells with nanoscale control of the interpenetrating network morphology," *Advanced Functional Materials*, vol. 15, no. 10, pp. 1617–1622, 2005.
2. G. Li, V. Shrotriya, J. Huang, Y. Yao, T. Moriarty, K. Emery, and Y. Yang, "High-efficiency solution processable polymer photovoltaic cells by self-organization of polymer blends," *Nature Materials*, vol. 4, no. 11, pp. 864–868, 2005.
3. http://www.ecnmag.com/news/2011/12/Heliatek-Organic-Solar-Cells-Achieve-Record-Efficiency/.
4. B. A. Gregg, and M. C. Hanna, "Comparing organic to inorganic photovoltaic cells: Theory, experiment, and simulation," *Journal of Applied Physics*, vol. 93, no. 6, pp. 3605–3614, 2003.
5. J. J. M. Halls, C. A. Walsh, N. C. Greenham, E. A. Marseglia, R. H. Friend, S. C. Moratti, and A. B. Holmes, "Efficient photodiodes from interpenetrating polymer networks," *Nature*, vol. 376, no. 6540, pp. 498–500, 1995.
6. G. Yu, J. Gao, J. C. Hummelen, F. Wudl, and A. J. Heeger, "Polymer photovoltaic cells: Enhanced efficiencies via a network of internal donor-acceptor heterojunctions," *Science*, vol. 270, no. 5243, pp. 1789–1791, 1995.
7. D. Carsten, and D. Vladimir, "Polymer–fullerene bulk heterojunction solar cells," *Reports on Progress in Physics*, vol 73, no. 9, p. 096401, 2010.
8. C. J. Brabec, N. S. Sariciftci, and J. C. Hummelen, "Plastic solar cells," *Advanced Functional Materials*, vol. 11, no. 1, pp. 15–26, 2001.
9. T. M. Clarke, and J. R. Durrant, "Charge photogeneration in organic solar cells," *Chemical Reviews*, vol. 110, pp. 6736–6767, 2010.
10. X. Zhao, B. Mi, Z. Gao, and W. Huang, "Recent progress in the numerical modeling for organic thin film solar cells," *Science China Physics, Mechanics and Astronomy*, vol. 54, no. 3, pp. 375–387, 2011.
11. L. A. A. Pettersson, L. S. Roman, and O. Inganas, "Modeling photocurrent action spectra of photovoltaic devices based on organic thin films," *Journal of Applied Physics*, vol. 86, no. 1, pp. 487–496, 1999.

12. D. W. Sievers, V. Shrotriya, and Y. Yang, "Modeling optical effects and thickness dependent current in polymer bulk-heterojunction solar cells," *Journal of Applied Physics,* vol. 100, no. 11, pp. 114509–7, 2006.
13. R. Hausermann, E. Knapp, M. Moos, N. A. Reinke, T. Flatz, and B. Ruhstaller, "Coupled optoelectronic simulation of organic bulk-heterojunction solar cells: Parameter extraction and sensitivity analysis," *Journal of Applied Physics,* vol. 106, no. 10, pp. 104507–9, 2009.
14. V. D. Mihailetchi, L. J. A. Koster, J. C. Hummelen, and P. W. M. Blom, "Photocurrent generation in polymer-fullerene bulk heterojunctions," *Physical Review Letters,* vol. 93, no. 21, p. 216601, 2004.
15. X. N. Yang, J. Loos, S. C. Veenstra, W. J. H. Verhees, M. M. Wienk, J. M. Kroon, M. A. J. Michels, and R. A. J. Janssen, "Nanoscale morphology of high-performance polymer solar cells," *Nano Letters,* vol. 5, no. 4, pp. 579–583, 2005.
16. L. J. A. Koster, "Charge carrier mobility in disordered organic blends for photovoltaics," *Physical Review B,* vol. 81, no. 20, p. 205318, 2010.
17. F. Xu, and D. Yan, "The role of mobility in bulk heterojunction solar cells," *Applied Physics Letters,* vol. 99, no. 11, p. 113303, 2011.
18. A. Cravino, "Origin of the open circuit voltage of donor-acceptor solar cells: Do polaronic energy levels play a role?," *Applied Physics Letters,* vol. 91, no. 24, p. 243502, 2007.
19. K. Vandewal, K. Tvingstedt, A. Gadisa, O. Inganas, and J. V. Manca, "On the origin of the open-circuit voltage of polymer-fullerene solar cells," *Nature Material,* vol. 8, no. 11, pp. 904–909, 2009.
20. C. J. Brabec, A. Cravino, D. Meissner, N. S. Sariciftci, T. Fromherz, M. T. Rispens, L. Sanchez, and J. C. Hummelen, "Origin of the open circuit voltage of plastic solar cells," *Advanced Functional Materials,* vol. 11, no. 5, pp. 374–380, 2001.
21. L. J. A. Koster, E. C. P. Smits, V. D. Mihailetchi, and P. W. M. Blom, "Device model for the operation of polymer/fullerene bulk heterojunction solar cells," *Physical Review B,* vol. 72, no. 8, p. 085205, 2005.
22. Q. Gu, E. A. Schiff, S. Grebner, F. Wang, and R. Schwarz, "Non-Gaussian Transport Measurements and the Einstein Relation in Amorphous Silicon," *Physical Review Letters,* vol. 76, no. 17, p. 3196, 1996.
23. Y. Roichman, and N. Tessler, "Generalized Einstein relation for disordered semiconductors—implications for device performance," *Applied Physics Letters,* vol. 80, no. 11, pp. 1948–1950, 2002.
24. H. J. Snaith, L. Schmidt-Mende, M. Gratzel, and M. Chiesa, "Light intensity, temperature, and thickness dependence of the open-circuit voltage in solid-state dye-sensitized solar cells," *Physical Review B,* vol. 74, no. 4, p. 045306, 2006.
25. R. Yohai, and T. Nir, "Generalized Einstein relation for disordered semiconductors—implications for device performance," *Applied Physics Letters,* vol. 80, no. 11, pp. 1948–1950, 2002.
26. G. Juska, K. Genevicius, N. Nekrasas, G. Sliauzys, and R. Osterbacka, "Two dimensional Langevin recombination in regioregular poly(3-hexylthiophene)," *Applied Physics Letters,* vol. 95, no. 1, p. 013303, 2009.
27. G. Juska, K. Genevicius, N. Nekrasas, and G. Sliauzys, "Two-dimensional Langevin recombination," *Physica Status Solidi (c),* vol. 7, no. 3–4, pp. 980–983, 2010.

28. C. Deibel, A. Baumann, and V. Dyakonov, "Polaron recombination in pristine and annealed bulk heterojunction solar cells," *Applied Physics Letters,* vol. 93, no. 16, pp. 163303–3, 2008.
29. G. Juška, K. Genevičius, N. Nekrašas, G. Sliaužys, and G. Dennler, "Trimolecular recombination in polythiophene: fullerene bulk heterojunction solar cells," *Applied Physics Letters,* vol. 93, no. 14, p. 143303, 2008.
30. G. A. H. Wetzelaer, M. Kuik, M. Lenes, and P. W. M. Blom, "Origin of the dark-current ideality factor in polymer:fullerene bulk heterojunction solar cells," *Applied Physics Letters,* vol. 99, no. 15, p. 153506, 2011.
31. C. G. Shuttle, R. Hamilton, B. C. O'Regan, J. Nelson, and J. R. Durrant, "Charge-density-based analysis of the current–voltage response of polythiophene/fullerene photovoltaic devices," *Proceedings of the National Academy of Sciences USA,* vol. 107, no. 38, pp. 16448–16452, 2010.
32. C. R. McNeill, S. Westenhoff, C. Groves, R. H. Friend, and N. C. Greenham, "Influence of Nanoscale Phase Separation on the Charge Generation Dynamics and Photovoltaic Performance of Conjugated Polymer Blends: Balancing Charge Generation and Separation," *Journal of Physical Chemistry C,* vol. 111, no. 51, pp. 19153–19160, 2007.
33. R. A. Street, M. Schoendorf, A. Roy, and J. H. Lee, "Interface state recombination in organic solar cells," *Physical Review B,* vol. 81, no. 20, p. 205307, 2010.
34. L. Liu, and G. Li, "Investigation of recombination loss in organic solar cells by simulating intensity-dependent current-voltage measurement," *Solar Energy Materials and Solar Cells,* vol. 95, no. 9, pp. 2557–2563, 2011.
35. S. R. Cowan, A. Roy, and A. J. Heeger, "Recombination in polymer-fullerene bulk heterojunction solar cells," *Physical Review B,* vol. 82, no. 24, p. 245207, 2010.
36. L. Liu, and G. Li, "Modeling and simulation of organic solar cells," in *IEEE Nanotechnology Materials and Devices Conference,* Monterey, California, 2010.
37. D. L. Scharfetter, and H. K. Gummel, "Large-signal analysis of a silicon Read diode oscillator," *IEEE Transactions on Electron Devices,* vol. 16, no. 1, pp. 64–77, 1969.
38. H. K. Gummel, "A self-consistent iterative scheme for one-dimensional steady state transistor calculations," *IEEE Transactions on Electron Devices,* vol. 11, no. 10, pp. 455–465, 1964.
39. A. M. Winslow, "Numerical solution of the quasilinear poisson equation in a nonuniform triangle mesh," *Journal of Computational Physics,* vol. 1, no. 2, pp. 149–172, 1966.
40. E. Knapp, R. Hausermann, H. U. Schwarzenbach, and B. Ruhstaller, "Numerical simulation of charge transport in disordered organic semiconductor devices," *Journal of Applied Physics,* p. 054504, 2010.
41. X. Yang, J. Loos, S. C. Veestra, W. J. H. Verhees, M. M. Wienk, J. M. Kroon, M. A. J. Michels, and R. A. J. Janssen, "Nanoscale morphology of high-performance polymer solar cells," *Nano Letters,* vol. 5, no. 4, pp. 579–583, 2005.
42. H. Hoppe, M. Niggemann, C. Winder, J. Kraut, R. Hiesgen, A. Hinsch, D. Meissner, and N. S. Sariciftci, "Nanoscale morphology of conjugated polymer/fullerene-based bulk-heterojunction solar cells," *Advanced Functional Materials,* vol. 14, no. 10, pp. 1005–1011, 2004.

43. Y. Zhao, Z. Xie, Y. Qu, Y. Geng, and L. Wang, "Solvent-vapor treatment induced performance enhancement of poly(3-hexylthiophene):methanofullerene bulk-heterojunction photovoltaic cells," *Applied Physics Letters,* vol. 90, no. 4, p. 043504, 2007.
44. S. Miller, G. Fanchini, Y.-Y. Lin, C. Li, C.-W. Chen, W.-F. Su, and M. Chhowalla, "Investigation of nanoscale morphological changes in organic photovoltaics during solvent vapor annealing," *Journal of Materials Chemistry,* vol. 18, no. 3, pp. 306–312, 2008.
45. S. G. Brush, "History of the Lenz-Ising model," *Reviews of Modern Physics,* vol. 39, no. 4, pp. 883–893, 1967.
46. M. Casalegno, G. Raos, and R. Po, "Methodological assessment of kinetic Monte Carlo simulations of organic photovoltaic devices: The treatment of electrostatic interactions," *Journal of Chemical Physics,* vol. 132, no. 9, p. 094705, 2010.
47. F. Yang, and S. R. Forrest, "Photocurrent generation in nanostructured organic solar cells," *ACS Nano,* vol. 2, no. 5, pp. 1022–1032, 2008.
48. R. A. Marsh, C. Groves, and N. C. Greenham, "A microscopic model for the behavior of nanostructured organic photovoltaic devices," *Journal of Applied Physics,* vol. 101, no. 8, p. 083509, 2007.

4

Optimization of Organic Photovoltaic Cells

Fanan Wei, Liming Liu, and Guangyong Li

CONTENTS

4.1 Introduction

With the advantages of being low cost, lightweight and flexible, organic photovoltaics are becoming a promising technology for sustainable energy harvesting. As a result of the continuous efforts of the worldwide research community, the current power conversion efficiency (PCE) record of organic photovoltaic cells has reached 12% [1] and the projected lifetime is more than 7 years [2]. Nevertheless, almost all advancements in organic photovoltaics so far have been from trial-and-error experiments. Because of the complexity of organic photovoltaic cells, it is not possible to simply tune one or a few parameters to achieve optimal performance. Further advancement in organic photovoltaics will be a daunting challenge, considering the nearly infinite number of material combinations and processing conditions that can affect device performance. Compared to trial-and-error-based physical experiments, computational modeling and simulation is a more effective and economical way to optimize devices for best performance. In Chapter 3, we described the optical and electrical modeling and simulation of organic photovoltaic cells. If these approaches can be combined to include all aspects of organic photovoltaic cells in the model, optimization of organic photovoltaic cells will be possible. In this chapter, we describe how organic photovoltaic cells are optimized for best performance via simulations. The poly-3-hexylthiophene (P3HT):phenyl-C61-butyric acid methylester (PCBM) material system is used as an example to demonstrate the effectiveness of optimization via simulation. The P3HT:PCBM system is chosen simply because it is commonly investigated in the literature and many parameters are readily available. The methods described herein could be applied to any other type of organic photovoltaic cells without any technical barriers.

4.2 Optimizing Device Thickness via Optical Modeling and Electrical Simulation

Carrier mobility is one of the important factors that dictate the performance of the photovoltaic cells. It has been found that carrier mobility in organic photovoltaic cells can be improved by ameliorating the nanoscale morphology of the active layer through thermal annealing [3–6] and solvent annealing [6–8]. The improved carrier mobility is largely due to the improved crystallinity and phase separation, which has been proven with XRD (x-ray diffraction) and TEM [3]. Table 4.1 shows the typical carrier mobility collected from the literature under different annealing conditions. Clearly, carrier mobility, especially hole mobility, can be significantly improved through thermal annealing. These parameters can be used to predict device performance via simulation.

In Chapter 3, we developed an optical model to calculate the photon absorption of organic photovoltaic cells based on optical transfer matrix. By taking the interference effect into consideration, the optical transfer matrix can be used to compute the optical electric field and subsequently light intensity distribution in organic photovoltaic cells. By integrating the energy dissipation rate for light wavelength between 350 nm and 800 nm (effective absorption range for P3HT), the photon absorption efficiency under standard AM 1.5G illumination with respect to the thickness of the P3HT:PCBM layer for a device with the structure ITO(150 nm)/PEDOT:PSS(40 nm)/P3HT:PCBM-/Al(100 nm) can be acquired as shown in Figure 4.1. Clearly, there exist absorption peaks due to the interference, which means the maximum power conversion efficacy will be very sensitive to the thickness of the active layer.

By combining optical modeling and electrical simulation using drift-diffusion model as described in Chapter 3, we are able to optimize the

TABLE 4.1

Carrier Mobility for Figure 4.2 Simulation

Mobility	Device	As-cast	Annealed at 70°C	Annealed at 140°C
μ_n (cm^2/Vs)		2×10^{-4}	1×10^{-3}	2.5×10^{-3}
μ_p (cm^2/Vs)		4×10^{-8}	1×10^{-6}	3×10^{-4}

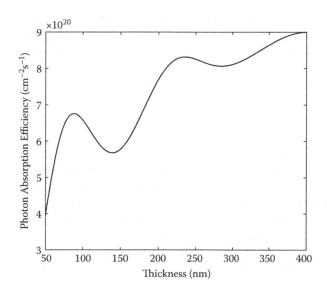

FIGURE 4.1
Photon absorption efficiency versus the thickness of the active layer.

thickness of organic photovoltaic cells by simulating *J–V* curves for devices with different thicknesses and then by extracting parameters such as short-circuit current density (J_{sc}), open-circuit voltage (V_{oc}), fill factor (*FF*), and power conversion efficiency (PCE) as a function of the thickness of the active layer. Figure 4.2 shows the simulation results for P3HT:PCBM solar cell

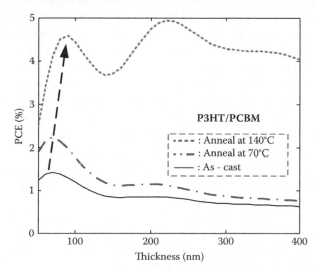

FIGURE 4.2
Thickness-dependent performance of P3HT/PCBM photovoltaic cells.

devices with different post-treatments: as-casted (solid curve), annealed at 70°C (dotted dash curve), and annealed at 140°C for one half hour (dotted curve), respectively. The PCE shown in Figure 4.2 gives different peak locations for different devices. There are two observations regarding the PCE in Figure 4.2: First, the PCE increases with the increase in carrier mobility (Table 4.1), irrespective of the thickness of the active layer; second, with the increased carrier mobility, the thickness value at which PCE gets maximum value increased as indicated by the arrow in the inset because high carrier mobility can guarantee sufficient carrier extraction when increasing the thickness. Compared with the other two devices, the increased performance of the device annealed at 140°C is mainly due to the increased carrier mobility [9,10]. Therefore, the location of the optimal thickness is also determined by the carrier mobility and hence this simulation is indispensable to optimizing the thickness of organic photovoltaic cells. The simulation results agree well with the results reported in the literature that the best device performance is located at either thickness around 90 nm or 220 nm.

4.3 Optimizing Device via Multiscale Simulation

Nanoscale morphology of the active layer is the main factor that dictates carrier mobility, and thus device performance. The optimization in the last section combines only the optical model and the electrical simulation based on a drift and diffusion model, and thus is unable to directly take nanoscale morphology into account. A multiscale simulation strategy that integrates Monte Carlo simulation with optical absorption calculation and macroscopic simulation will be a suitable approach to simulate the performance of organic photovoltaic cells. Such integration takes into account the multiscale aspects of the device such as the active layer thickness, nanoscale morphology, and the weight ratio of donor/acceptor. Thus, it is effective in optimizing several physical aspects of the device simultaneously. In this section, we describe the details of optimizing the device via this multiscale simulation.

The schematic diagram of the proposed multiscale modeling and simulation framework, illustrated in Figure 4.3, is composed of four modules:

1. *Morphology Generation Module*: The morphology generation module constructs the nanoscale morphology of the bulk heterojunctions according to the weight ratio of D/A as well as other constraints and parameters such as annealing conditions.

2. *Monte Carlo Simulation Module*: The Monte Carlo simulation module calculates the recombination ratio, exciton-dissociation efficiency,

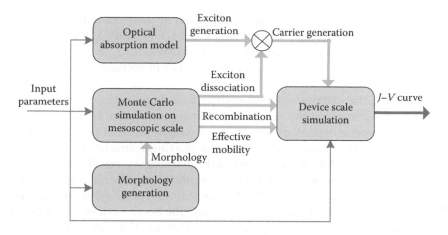

FIGURE 4.3
Schematic of multiscale simulation of organic photovoltaic cells with bulk heterojunctions.

and the bulk (apparent or effective) mobility corresponding to the internal morphology and the internal electrical field.

3. *Carrier Generation Calculation Module*: The carrier generation calculation module obtains the free charge carrier generation rate by multiplying the exciton-dissociation efficiency (EDE) obtained from the Monte Carlo simulation and the light absorption efficiency calculated from an optical absorption model based on the active layer thickness and the D/A weight ratio.

4. *Macroscopic Simulation Module*: Those intermediate parameters obtained from previous modules are fed into the macroscopic simulation module that predicts the current–voltage responses (*J–V* curves) of a device by considering the physical dimension and boundary conditions as well as other properties of the device. The performances of the device such as the short-circuit current, the open-circuit voltage, the fill factor, and the power conversion efficiency (PCE) are derived from these *J–V* curves.

During the Monte Carlo simulation, the trajectories of particles (exciton, electron, and hole) are completely recorded for postanalysis. Several variables can be estimated when the simulation reaches a stable state. The effective carrier mobility (bulk mobility), the exciton EDE, and the recombination rate are calculated from the following equations:

$$\mu_{effc} = l_{field} \Big/ \Big(F \sum \tau \Big) \tag{4.1}$$

$$EDE = n_{ex\_diss} / n_{exciton} \tag{4.2}$$

$$Rr = n_{recom\_pair} / n_{ex\_diss} \tag{4.3}$$

where l_{field} is hopping distance in the direction of the electric field, τ is total time expired for hopping, F is the electric field, $F = (\Delta\phi - V_{ext})/d$, in which V_{ext} is the applied external voltage, and d is thickness; n_{ex_diss} is the number of dissociated excitons, and n_{recom_pair} is the number of recombined charge carrier pairs.

By varying the external applied voltage and domain size (the size of the minority component) of the active layer, the dependence of bulk mobility on electric field and morphology for different weight ratios of the P3HT:PCBM active layer can be determined from the Monte Carlo simulation. To improve the accuracy, each data point is obtained by averaging results from at least three repeated simulations under the same conditions. The simulation results demonstrate that the effective (bulk) mobility increases when the domain size increases (Figures 4.4a and 4.5a) because the barrier is less for carriers to move toward the corresponding electrode. The effective mobility decreases when the electric field increases (Figures 4.4b and 4.5b) because it is more difficult for carriers to escape from dead-end traps under high electric field strength. The result agrees well with that from the experiment [11]. It is also clear that when the concentration of PCBM is higher, a higher electron mobility is observed, while a higher concentration of P3HT leads to a higher hole mobility, which is intuitive and straightforward.

The recombination rates corresponding to the electric field and morphology for different weight ratios are shown in Figure 4.6. The recombination rate shows a declining tendency to both domain size and electric field strength. As the domain size increases, the transport paths for carriers become wider, and the carriers are more likely to travel to the electrodes rather than being recombined at the donor/acceptor interface. Under a stronger electric field, the carriers can move to electrodes more quickly, so the recombination rate is

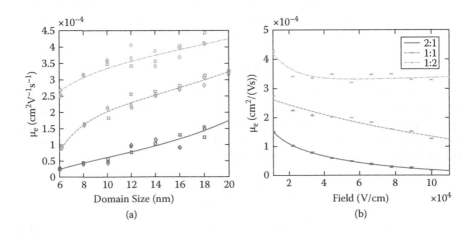

(a)   (b)

FIGURE 4.4

Dependences of effective electron mobility (bulk electron mobility) on (a) domain size and (b) electric field. The legend shows the corresponding D/A weight ratio.

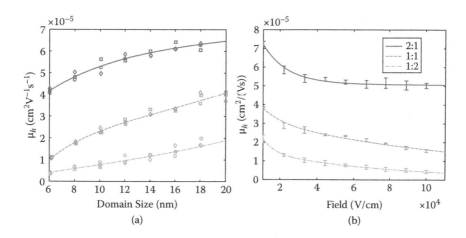

FIGURE 4.5
Dependencies of effective hole mobility (bulk hole mobility) on (a) domain size and (b) field. The legend shows the corresponding D/A weight ratio.

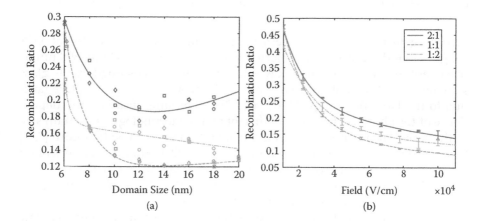

FIGURE 4.6
Dependencies of the recombination ratio on (a) domain size and (b) field. The legend shows the corresponding D/A weight ratio.

expected to decrease when the electric field strength increases. It is observed that the 2:3 weight ratio gives the least recombination rate.

The dependence of EDE on domain size is also calculated in the simulation (Figure 4.7). It can be seen that the EDE decreases as domain size increases. Such a trend is reasonable because larger domain size means less interfacial area and thus lower EDE. Apparently, the 1:1 ratio yields the best EDE, which is the reason why the 1:1 ratio often gives the best experimental results.

The photon absorption rate can be computed through integration of the energy dissipation rate for light wavelength between 350 nm and 800 nm

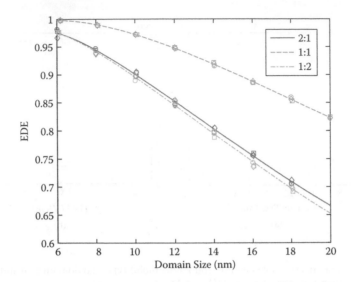

FIGURE 4.7
Dependencies of EDE on the domain size of the active layer. The weight ratio is shown in the legend.

according to the extinction coefficients for materials with different D/A weight ratios (Figure 4.8). The selected wavelength range (350 nm–800 nm) is reasonable because the light below 350 nm is strongly absorbed by the glass substrate, and the light above 800 nm is rarely absorbed by an active layer due to the large band gap (1.9 eV) of P3HT. For simplification, the dependence of the photon extinction coefficient on the morphology or domain size of the active layer is not taken into consideration here. Some literature [8,10] has reported that the absorption profile changes little after device annealing even though the inner morphology and external quantum efficiency change dramatically. Therefore, it is fair to assume that the photon absorption coefficient remains constant no matter the domain size.

Based on the optical transfer matrix theory [11], we are able to calculate the absorption efficiency of the P3HT:PCBM active layer according to the thickness and D/A weight ratio. Herein, the illumination intensity is set to be the AM 1.5G spectrum. From a Monte Carlo simulation, we have obtained the EDE with respect to domain size (Figure 4.7). If we assume that one absorbed photon transfers into one exciton, the charge carrier generation rate (Figure 4.9) can be obtained now by multiplying the EDE and the photon absorption efficiency.

Even though the internal quantum efficiency, *J–V* curve and subsequently PCE are obtainable from Monte Carlo simulation directly, Monte Carlo simulation is not a proper choice for a full device description because it does not have a full consideration of the boundary conditions at the electrodes–active layer interfaces. In addition, the Monte Carlo simulation of the *J–V* curve has

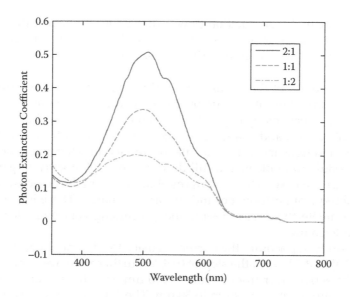

FIGURE 4.8
Photon extinction coefficient for different D/A weight ratio of P3HT:PCBM.

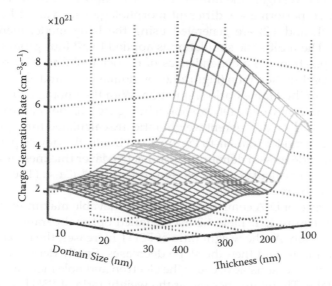

FIGURE 4.9
The dependence of the charge carrier generation rate on both domain size and thickness of an active layer for a device with D/A weight ratio 1:1.

a demand for high computation power that is often not readily available. To obtain the *J–V* curve of the organic photovoltaic cells, we use the macroscopic simulation based on the drift-diffusion model that was discussed in Chapter 3.

After integration of the simulation modules at each scale, we are able to perform optimization of organic photovoltaic cells via the multiscale simulation. The morphology-dependent parameters, such as bulk mobility, recombination rate, and charge generation rate obtained from Monte Carlo simulation and light absorption calculation, are fed into the macroscopic simulation module to implement a full description of photocurrent generation on the device scale. With the *J–V* curves obtained from the device scale simulation, subsequent performance indices are calculated. These indices can be used to evaluate the device design and processing conditions for optimal device performance.

In this study, we acquire the current density for the input applied voltage from −0.3 V to 0.7 V with the step of 0.01 V. Furthermore, we vary the thickness of the active layer from 50 nm to 400 nm and the domain size (of the minority component) from 4 nm to 30 nm. The *J–V* curves corresponding to each point of thickness and domain size and the subsequent device performance parameters are all calculated. As performance dependence on morphology and thickness is predicted, we can locate the maximum PCE value and its corresponding domain size and thickness. Thus, both the inner morphology and the external physical dimension are optimized at the same time.

In order to investigate the influence of the composition of the active layer on the device performance, different morphologies for varied D/A weight ratio (2:1, 1:1, and 1:2) are generated using the Ising model mentioned in Section 3.4. The multiscale simulation is applied to all four groups with different compositions. The dependences of carrier mobility (Figures 4.4 and 4.5) and recombination ratio (Figure 4.6) on domain size and electric field are obtained from the Monte Carlo simulation. With the photon extinction coefficient for different weight ratios, shown in Figure 4.8, the charge generation rates corresponding to different weight ratio are calculated through the optical absorption model (similar to Figure 4.1). Finally, the device performance indices' dependences on domain size and active layer thickness for different weight ratios are calculated and plotted in Figures 4.10a–c. (The results of a 3:1 weight ratio are not shown.)

The influence of D/A composition on the achievable maximum PCE value is shown in Figure 4.11. Table 4.2 lists detailed optimal parameters for each D/A composition. As the weight ratio of P3HT increases, the maximum PCE increases initially, and then the slope decreases, finally decreasing quickly. The tendency is reasonable because the electron and hole mobility are unbalanced initially. Therefore, increasing the weight ratio of P3HT increases the hole mobility and balance of the transport of the two types of carriers. Further, the increase in P3HT composition will lead to the rise of photon absorption efficiency, and result in a higher maximum PCE. Excessively increasing

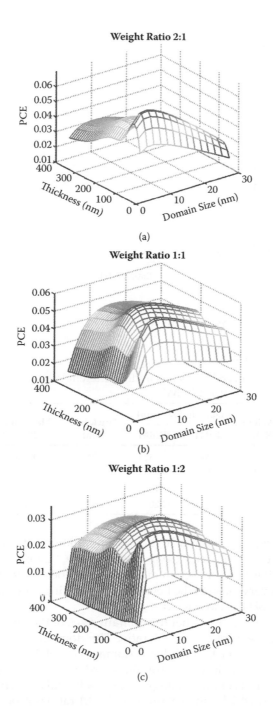

FIGURE 4.10

The dependence of PCE on the domain size and thickness of an active layer for varied D/A composition: (a) 2:1, (b) 1:1, and (c) 1:2.

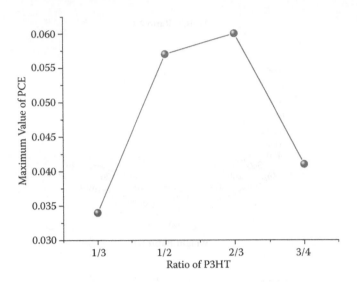

FIGURE 4.11
The maximum value of PCE versus weight ratio of P3HT.

TABLE 4.2

Optimized Parameters under Different Weight
Ratios of P3HT:PCBM

Weight Ratio	Domain Size (nm)		Thickness (nm)	Maximum PCE
	P3HT	**PCBM**		
3:1	24	8	80	4.1%
2:1	16	8	80	6.0%
1:1	10	10	90	5.7%
1:2	10	20	80	3.4%

P3HT composition results in a higher recombination rate (Figure 4.6) and thus leads to a dramatic decrease in performance. Surprisingly, the weight ratio 1:1, which has been chosen in many studies, does not yield the best performance but is very close to the best one. Our findings suggest that the best weight ratio of P3HT is close to 2:1.

From Table 4.2, we can see that the domain sizes of the donor and acceptor are asymmetrical except for the 1:1 ratio. The domain size of the minority component remains constant while the domain size of the counterpart needs to grow with respect to its concentration in order to achieve best device performance. Such kinds of morphology raise a technical challenge to be implemented in reality, which can explain why the 1:1 ratio is reported most often in the literature.

4.4 Discussion and Conclusion

In summary, we first optimized the device thickness by combining the optical model and the electrical simulation based on the drift-diffusion model. We then proposed a framework for multiscale modeling and simulation of organic photovoltaic cells with bulk heterojunctions by integrating the Monte Carlo simulation, the photon absorption calculation based on optical transfer matrix theory, and the device scale simulation based on the drift-diffusion model. The multiscale approach is able to account for multiscale aspects of the bulk heterojunctions in the simulation. In this integration, the Monte Carlo simulation engenders exciton-dissociation efficiency, bulk mobility, and recombination rate by taking into account the nanoscale morphology; the outputs of the Monte Carlo simulation are then rendered to the light absorption calculator and the device scale simulator that further generates the complete current-voltage (J–V) characteristics according to the physical dimensions, boundary conditions, and other parameters of the device. The proposed multiscale simulation method can be an efficient tool to simultaneously optimize the thickness, morphology, and D/A composition of the bulk heterojunctions. Even though some parameters used in this simulation may not be completely accurate, the optimization results from such a multiscale simulation approach will provide useful guidance in optimizing the real device through experiments, thus saving a tremendous amount of effort in searching for the optimal parameters. The multiscale simulation is demonstrated for a P3HT:PCBM system in this chapter, but it can be easily extended to other types of organic bulk heterojunctions solar cells as long as the properties of the materials are known.

References

1. http://www.pv-tech.org/news/verified_heliatek_organic_solar_cell_achieves_record_12 conversion efficien.
2. C. H. Peters, I. T. Sachs-Quintana, J. P. Kastrop, S. Beaupré, M. Leclerc, and M. D. McGehee, "High efficiency polymer solar cells with long operating lifetimes," *Advanced Energy Materials*, vol. 1, no. 4, pp. 491–494, 2011.
3. W. L. Ma, C. Y. Yang, X. Gong, K. Lee, and A. J. Heeger, "Thermally stable, efficient polymer solar cells with nanoscale control of the interpenetrating network morphology," *Advanced Functional Materials*, vol. 15, no. 10, pp. 1617–1622, 2005.
4. X. Yang, J. Loos, S. C. Veestra, W. J. H. Verhees, M. M. Wienk, J. M. Kroon, M. A. J. Michels, and R. A. J. Janssen, "Nanoscale morphology of high-performance polymer solar cells," *Nano Letters*, vol. 5, no. 4, pp. 579–583, 2005.

5. H. Hoppe, M. Niggemann, C. Winder, J. Kraut, R. Hiesgen, A. Hinsch, D. Meissner, and N. S. Sariciftci, "Nanoscale morphology of conjugated polymer/fullerene-based bulk-heterojunction solar cells," *Advanced Functional Materials*, vol. 14, no. 10, pp. 1005–1011, 2004.

6. Y. Zhao, Z. Xie, Y. Qu, Y. Geng, and L. Wang, "Solvent-vapor treatment induced performance enhancement of poly(3-hexylthiophene): Methanofullerene bulk-heterojunction photovoltaic cells," *Applied Physics Letters*, vol. 90, no. 4, p. 043504, 2007.

7. G. Li, V. Shrotriya, J. Huang, Y. Yao, T. Moriarty, K. Emery, and Y. Yang, "High-efficiency solution processable polymer photovoltaic cells by self-organization of polymer blends," *Nature Materials*, vol. 4, no. 11, pp. 864–868, 2005.

8. S. Miller, G. Fanchini, Y.-Y. Lin, C. Li, C.-W. Chen, W.-F. Su, and M. Chhowalla, "Investigation of nanoscale morphological changes in organic photovoltaics during solvent vapor annealing," *Journal of Materials Chemistry*, vol. 18, no. 3, pp. 306–312, 2008.

9. L. A. A. Pettersson, L. S. Roman, and O. Inganas, "Modeling photocurrent action spectra of photovoltaic devices based on organic thin films," *Journal of Applied Physics*, vol. 86, no. 1, pp. 487–496, 1999.

10. H. Yan, B. A. Collins, E. Gann, C. Wang, H. Ade, and C. R. McNeill, "Correlating the efficiency and nanomorphology of polymer blend solar cells utilizing resonant soft x-ray scattering," *ACS Nano*, vol. 6, no. 1, pp. 677–688, 2011.

11. S. Albrecht, W. Schindler, J. Kurpiers, J. Kniepert, J. C. Blakesley, I. Dumsch, S. Allard, K. Fostiropoulos, U. Scherf, and D. Neher, "On the field dependence of free charge carrier generation and recombination in blends of PCPDTBT/PC70BM: Influence of solvent additives," *Journal of Physical Chemistry Letters*, vol. 3, no. 5, pp. 640–645, 2012.

5

Developing a Dynamics Model for Epidermal Growth Factor (EGF)-Induced Cellular Signaling Events

Ning Xi, Ruiguo Yang, Bo Song, King Wai Chiu Lai, Hongzhi Chen, Jennifer Y. Chen, Lynn S. Penn, and Jun Xi

CONTENTS

5.1 Introduction

Epidermal growth factor (EGF) is a growth factor that binds a cell surface receptor [1] to an epidermal growth factor receptor (EGFR), to promote cell growth and proliferation. The ligand binding can trigger a series of downstream signaling events that regulate cell behaviors, resulting in cell proliferation, differentiation, or migration. These different cell biological behaviors result in different mechanical behaviors. The A431 human epidermoid carcinoma cell line has been widely used as a biochemical model for the investigation of signaling pathway–related cell responses triggered by EGF stimulation [2]. Researchers [3] have reported that EGF-treated A431 cells show significant rounding and swelling. It has long been postulated that signaling pathways could regulate the configuration of the cytoskeleton as a mechanism of facilitating the prospective cell behavior [4]. For instance, a cancer cell would display loose attachment, that is, poor adhesion, to make way for metastasis [5]; migration cells would have stronger cell adhesion in the filopodia in the leading edges of migration than in the trailing edges [6]. The configuration of the cytoskeleton thus will determine the overall structure of the cell body, and one can use the cytoskeleton configuration to reveal the biological behavior of living cells after certain biochemical cues. Since the cytoskeleton is deemed as the main force bearer of the whole cell [7,8], it can then be probed by mechanical force; the mechanical response of the cell, or cytoskeleton would provide abundant information about the mechanical status of the cell. We normally refer to these as biomechanical markers. Nanomechanical sensors are normally utilized to perform force loading and measurement; the measurement result is expected to yield unique insight into the cell behavior and cell signaling pathways.

Atomic force microscopy (AFM) is a high-resolution imaging technique [9]. The imaging modality is based on force interactions at the molecular level between a cantilever with a sharp tip attached at the free end and the sample beneath it. The force interaction is mainly determined by the distance between the two; thus by controlling the interaction force to a constant level, one can build a topography map of the sample by raster-scanning its entire surface. The nature of the measurement process and its ability to work in liquid environments make AFM an ideal tool for visualization of cellular and molecular structures in situ. Meanwhile, the probing interaction also provides the mechanical property of the cell samples [10]. We have observed ultrastructural changes in real time at the nanometer scale in cellular adhesions on keratinocyte cells [11]. Others have also reported results from AFM imaging for cellular structural observation, either on animal cell lines or bacterial cells [12]. Cellular elasticity can be measured by recording the force–displacement curves during AFM nanoindentation. Its effectiveness in providing biomarkers for physiological conditions has been validated previously [5,13]. Cellular viscoelasticity can be obtained by fitting AFM force–displacement curves to the Johnson–Kendall–Roberts (JKR) dynamic model [14].

The quartz crystal microbalance with dissipation (QCM-D) is another nanomechanical sensor that can measure interactions at the molecular level [15]. It has been used to investigate molecular processes, such as membrane formations [16], protein adsorptions [17], and cell spreading [18]. It operates in thickness-shear mode by oscillating a quartz crystal disk with high frequency (normally 5 MHz) and low amplitude (less than 1 nm in the lateral direction). It collects the frequency and amplitude changes of the quartz crystal and reveals the mechanical/structural status of the sample attached to the top of the sensor disk. The shift in resonance frequency indicates the mass adsorption or loss, while the change in energy dissipation alteration can be used to derive the viscoelasticity properties of the adsorbed film on the disk [19].

The combination of AFM and QCM-D for biological investigations has been described in previous works [20–22]. Functionally, QCM-D measures viscoelastic property changes in biological matters at different physiological conditions. The mechanical information can be verified by probing the sample surface topography with AFM. This combination was demonstrated in an earlier study [20] where the effect of different substrates on cell shape and cell mechanics was investigated. The topography and mechanical properties of the cells were obtained by AFM imaging and force measurement, and the QCM-D characterizes the adhesion between the cells and the substrates through frequency and dissipation monitoring. Instrumentation combining AFM with a QCM-D substrate has been developed and described [21]. The described system is equipped with an oscillating quartz crystal as the substrate for the AFM experiment, and the AFM imaging and QCM-D recording can be performed simultaneously on the same sample. However, attention to comparative characterization of cellular viscoelasticity by AFM and QCM-D has been lacking. It has been shown that AFM and QCM-D together have the capability for viscoelasticity measurement. While AFM scans and deforms cells from the top, QCM-D measures the sensor oscillation from the bottom where cell adhesion to the substrate dominates the acoustic wave decay. AFM performs cellular investigations as single cell analysis, and QCM-D studies the properties of a small population, a single cell layer spreading on the quartz crystal. In all these aspects, AFM and QCM-D can verify and complement each other.

5.1.1 AFM Energy Dissipation and Hysteresivity Measurements

The nanoindentation process by AFM consists of loading and unloading of the vertical force exerted by the cantilever as shown in Figure 5.1. The vertical piezo first drives the cantilever toward the cell sample and deforms it

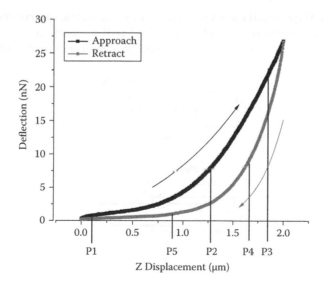

FIGURE 5.1
AFM energy dissipation.

$(P_1 \rightarrow P_2 \rightarrow P_3)$; then it retracts back to the original height $(P_3 \rightarrow P_4 \rightarrow P_5)$. During the process, cantilever deflection will be picked up by a position sensitive device (PSD) to generate the applied forces at each location, and this recorded relationship is the force–displacement curve. For each cycle, forces in the approach curve at each displacement position are always larger than those in the retraction curve. This hysteresis in applied force over the displacement defines the energy loss for the loading/unloading process. The energy required to deform the cell surface at a nominal distance on loading is not entirely recovered from the cell in withdrawing the same distance during unloading. If we define the energy under the approaching curve as A, and the energy under the retraction curve as R, the area within the hysteresis loop $(A - R)$ therefore represents the amount of energy dissipated during each cycle. Furthermore, we can define

$$\zeta = \frac{A}{A - R}$$

as the hysteresivity, a characteristic measure of the viscosity property [23]. This viscoelastic index ζ indicates the viscoelastic property of a material, with $\zeta = 0(A = R)$ indicating pure elastic material and $\zeta = 1(R = 0)$ indicating pure plastic material. Most materials will have ζ values $0 < \zeta < 1$, showing both viscous and elastic character [24].

5.1.2 QCM-D Energy Dissipation Measurement

Quartz crystal microbalance (QCM) uses a piezoelectric quartz crystal as an oscillator, and an alternating electrical field will generate a shear deformation of the surface [15], as shown in Figure 5.2a. The upper and lower surfaces

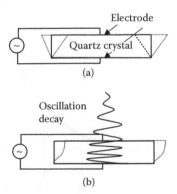

(a)

(b)

FIGURE 5.2
Schematic of QCM-D principle: (a) an alternating voltage will generate a relative movement between the upper and lower surfaces of the quartz crystal; (b) a decaying wave perpendicular to the sensor surface is the signature of QCM-D working in liquid.

will move in parallel but opposite directions laterally, generating an acoustic wave that will propagate in the direction perpendicular to the sensor disk surface. The acoustic wave has a frequency of $f = nv/(2t_q)$, where f is the resonant frequency, n is the overtone number, and t_q is the thickness of the crystal. When the QCM is operated in a gaseous environment, the oscillation decays negligibly. The Sauerbrey relationship describes a linear relationship between the mass change and the frequency shift [15]

$$\Delta m = -\frac{C}{n}\Delta f \tag{5.1}$$

where $C = t_q\rho_q/f_0$, and ρ_q is the density of the crystal.

However, when the sensor is immersed in liquid as shown in Figure 5.2b, the acoustic wave will propagate through the liquid and the oscillation amplitude will decay exponentially over time. The measurement circuit periodically switches the power input on and off, so that the free decay of oscillation can be obtained by recording the output voltage [15]. The voltage decay can be fitted by

$$A = A_0 e^{t/\tau}\sin(2\pi ft + \alpha) \tag{5.2}$$

The obtained parameters τ and f can then be used to evaluate the energy dissipation factor:

$$D = \frac{1}{\pi f\tau} \tag{5.3}$$

which can also be considered as the energy loss ratio of lost energy over stored energy. It is related to the viscoelasticity of the material film attached to the top of the sensor disk [25]. Several mechanical models have been proposed to relate the viscosity and elasticity of the film to the measured D value. Voinova and coworkers have adapted a continuum mechanics model that considers the oscillation wave propagating through a viscoelastic material. The material is immersed in a Newtonian fluid with either known or unknown properties [19], and the model produces the relationship of ΔD and Δf with the material properties on the top.

In the present work, we apply AFM probing and QCM-D observations to the same set of cellular signaling processes induced by EGF stimulation. Both experiments monitor a monolayer of A431 cells that grow on a substrate (silicate glass substrate for AFM and a quartz crystal surface for QCM-D) and are stimulated by EGF. While QCM-D monitors the basal area of the cell layer close to the quartz crystal surface where cell adhesion dominates, AFM probes the apical surface of a single cell away from the measurements-supporting surface, thereby revealing cytoplasm behavior.

By analyzing and comparing the quantitative data using both techniques, a more complete profile of the viscoelastic behavior of the cell can be obtained.

Once the mechanical profile is obtained from both measurements, a model of the signaling pathway dynamics following the EGF simulation can be built by considering the cell mechanical structure as the plant, which makes the readings as the output of the system, and the signaling pathway as the controller. We can address the signaling events from the perspective of systematic control and identify the pathway-related dynamics model structures and parameters.

5.2 Model Development

5.2.1 AFM Viscoelastic Characterization

The following assumption was made for the force–displacement-based viscoelastic modeling; the cellular material is represented as a Voigt element, a parallel configuration of spring (k) and dashpot (η), and there is only a repulsive force between the AFM tip and the cell. Then we define the vertical displacement of the Z piezo as d and the deformation of the cell surface as x (Figure 5.3). The spring constant of the cantilever is defined as k_c and its deflection is d_c.

For the retraction process, based on the Voigt element configuration, the dynamic force balance can be defined as

$$F = k(x_m - x) + \eta \dot{x} \tag{5.4}$$

where x_m, a constant, is the maximum vertical deformation of the cellular body from the approach process and x, a function of t, is the vertical contact point elevation due to the release of the cantilever pressure in the withdrawal process. The force can also be calculated by cantilever deflection as

$$F = k_c(d_{cm} - \Delta d_c) \tag{5.5}$$

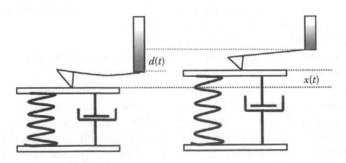

FIGURE 5.3
The modeling of viscoelastic material under AFM-based force-displacement measurement during the retraction process.

where d_{cm} is the maximum cantilever deflection resulting from the approach process and Δd_c is the release of deflection and is defined as the difference between the distance of retraction from the piezo (d) and the cell deformation recovery (x)

$$\Delta d_c = d - x \tag{5.6}$$

When Equations 5.5 and 5.6 are inserted into Equation 5.4, and one lets $d = vt$, where v is the velocity of the Z piezo motion, we have

$$\eta \dot{x} - (k_c + k)x = f_0 - k_c vt \tag{5.7}$$

where $f_0 = k_c d_{cm} - kx_m$ is a constant during the approach process. The solution to the ordinary differential equation, Equation 5.7, can be used to evaluate deformation recovery x under the influence of values η.

5.2.2 QCM-D-Based Cell Membrane Peeling Model

The sensor disk causes the relative motion of the cell with regard to the fluid in the sensor chamber. The effect is similar to that of the cell under liquid shear flow. In the leading edge, the adhesion bonds between the integrin receptors and the substrate ligands will be ruptured due to the flow-caused cell strain; at the other edge, there will be bond formation activities attributed to the increased integrin and ligand affinity, as shown in Figure 5.4. The model in Figure 5.4b describes this peeling of focal adhesion between the cell and the substrate due to shear stress under liquid flow. In the 2D model, focal adhesion sites are characterized by the contact length, L_c, between the cell and the substrate. The substrate moves with speed v_s, and the peeling speed is proportional to it according to $v_p = \kappa v_s$, where κ is the coefficient. Focal adhesion is modeled as a spring with a coefficient of k_a, whereas a single focal adhesion bond strength is $f_a = k_a(l - \lambda)$, where l and λ are the lengths of the bond in the stretched and unstretched states, respectively. The total focal adhesion strength then becomes $F_a = N_a k_a(l - \lambda)$, where N_a is the adhesion bond density. The adhesion kinetic equation specifies the balance for the formation and dissociation of adhesion bonds that determine the temporal- and spatial-dependent bond density $N_a(t,s)$ [26,27]:

$$\frac{\partial N_a}{\partial t} = v_p \frac{\partial N_a}{\partial s} + k_f (N_l^0 - N_a)(N_r^0 - N_a) - k_r N_a \tag{5.8}$$

$$k_f = k_f^0 \exp \frac{k_{ts}(l - \lambda)^2}{2k_b T} \tag{5.9}$$

$$k_r = k_r^0 \exp \frac{k_f - k_{ts}(l - \lambda)^2}{2k_b T} \tag{5.10}$$

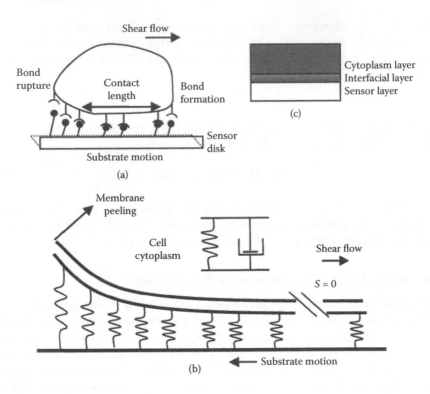

FIGURE 5.4
Cell membrane peeling model at the basal area (a, b); and the cell model as a whole (c) in which the cytoplasm is modeled as a Voigt element.

where N_l^0 is the ligand density on the substrate; N_r^0 is the cell surface receptor density; k_f is the bond formation rate with an initial value of k_f^0; k_r is the bond formation rate with an initial value k_r^0; and k_{ts} is the transient elastic constant of the adhesion bond. It is reasonable to assume that $k_{ts} < k_a$ for this model; k_b is the Boltzmann constant; and T is the temperature. For the calculation of the energy dissipation rate, we have

$$\frac{\partial N_a}{\partial t} = 0,$$

and Equations 5.8–5.10 can be evaluated numerically by the fourth-order Runge–Kutta method with the initial value N_a^0 solved by

$$0 = k_f(N_l^0 - N_a)(N_r^0 - N_a) - k_r N_a \tag{5.11}$$

The total energy dissipation rate for the focal adhesion dissociation can be defined as

$$E_a = w_c \int_0^{L_c} F_a v_p \frac{\partial l}{\partial s} \mathrm{d}x \tag{5.12}$$

where w_c is the effective width of the contact area.

5.3 Results and Discussion

The AFM measurement results before and after stimulation with EGF are shown in Figure 5.5a. The displacement was controlled below 2 μm, and the resultant loading force was within 30 nM, with the corresponding maximum cell deformation less than 500 nM (<25% of cell height). These parameters were used throughout the force measurement process. The hysteresis area

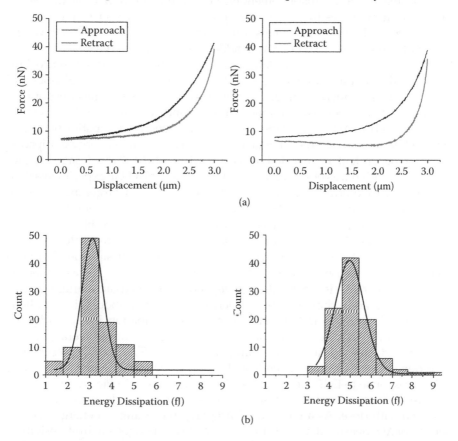

(a)

(b)

FIGURE 5.5
AFM energy dissipation measurement on 100 different cells with and without EGF treatment.

between the loading and unloading force curve $(A - R)$, was increased from 3.2 to 6.9 femto-Joules (fJ) due to the stimulation, a 126% jump.

A statistical analysis was performed on a population of cells before and after EGF stimulation. The result is shown in Figure 5.5b. The cell sample was first measured under AFM on 100 randomly selected cells. EGF was then applied to the Petri dish where the cell sample sat in a final concentration of 40 nM. The sample was then incubated for 30 min until another AFM measurement was conducted on 100 randomly selected cells. The plotted results showed a normal distribution for energy dissipation before and after the EGF addition. It can be observed that energy dissipation increased from 3.26 ± 0.95 fJ to 5.16 ± 0.89 fJ (mean $\pm$ SD). This demonstrates that the cells displayed different mechanical properties after EGF stimulation than before.

Given this general trend shown by the cell population, we want to define how cells dynamically adjust their cytoskeleton structure throughout the stimulation process. We monitored the stimulation process dynamically in real time. After EGF stimulation, force–displacement curves were taken at a 1 Hz frequency. Figure 5.6 summarizes the dynamic response of the cells after EGF treatment. Three different EGF concentrations were tested: 10 nM, 20 nM, and 40 nM. The same volume of buffer solution was added in a separate control experiment. The baseline measurement was longer at 40 min before EGF or the control buffer was added. For all three dosages, a few minutes after the administration of EGF, there was an increase in energy dissipation, suggesting the effect of EGF is immediate; this increase continued until around 20 min when the response started to level off. At the end of the observation, the energy dissipation was twice the baseline value. The experimental result also showed that energy dissipation was dose-dependent, as revealed by a comparison of the three dosages. Larger energy dissipation was observed for the 20 nM and 40 nM EGF experiments than for 10 nM EGF experiments. The 40 nM dose had a slightly larger increase than the 20 nM dose, although the difference was small (Figure 5.6). Similar dynamics were observed for the hysteresivity measurement (Figure 5.6b).

The results of QCM-D continuous monitoring for 100 min after stimulation with EGF of dosages confirmed that the energy dissipation decreases over time. Stimulation with EGF of different concentrations showed that the dissipation decrease is clearly dosage dependent. Immediately after stimulation, the dissipation showed a sharp increase, then after about a half an hour it dropped down. The initial sharp increase in dissipation is caused by the liquid flush during the change of EGF-containing medium. Overall, energy dissipation decreased due to the stimulation. Figure 5.7 shows representative dose-dependent and time-dependent changes in ΔD of the monolayer of A431 cells in response to EGF. Unlike the AFM results shown in Figure 5.6, where ζ increased with time and with dose, the QCM-D detected a decrease in ΔD with time and with dose. As a mechanical property, the change in energy dissipation factor ΔD measured directly by the QCM-D is analogous to hysteresivity, ζ, computed from the AFM measurements, where both quantities correspond to the mechanical energy loss relative to the energy input per measurement cycle.

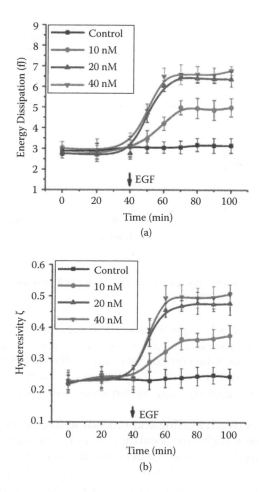

FIGURE 5.6
Dynamic mechanical responses. (a) Energy dissipation and (b) hysteresivity of A431 cells to stimulations with four different concentrations of EGF: 0, 10 nM, 20 nM, and 40 nM.

However, the two techniques probe different regions of the deposited cells; the AFM tip probes 500 nm into the cell from its top surface while the QCM-D senses approximately 100 nm up from the bottom surface of the sensor crystal.

Interestingly, responses for D and ζ reached their plateaus within a similar timeframe after the addition of the EGF (40–50 min for the decrease in ΔD and 50 min for the increase in ζ). Both changes showed signs of saturation in a similar dose range (responses at 20 nM and 40 nM at very similar levels). These similarities indicate that both mechanical responses (D and ζ) that were developed locally might be synchronously mediated by the same globe cell signaling pathway(s).

For pure elastic materials, there will be no energy dissipation. Accordingly, when an AFM force measurement is performed on a perfectly elastic material,

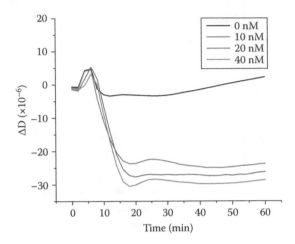

FIGURE 5.7
QCM-D measurement of energy dissipation over time after EGF treatment with concentrations of 0 nM, 10 nM, 20 nM, and 40 nM.

the approach and retraction portions of the force–displacement curve overlap. However, for biological samples, viscosity is inherent, and research has shown that internal cellular components, such as the cytoplasm and even the cytoskeleton, are viscoelastic [28]. Therefore, typical force curves on biological samples in liquid display large hysteresis between the approach and the retraction curves.

The hysteresis between loading and unloading force curves indicates that the repulsive force at each displacement position in unloading is lower than in loading. The origin of the decreased repulsion on the unloading curve is that at any given nominal displacement the actual local separation is larger on retraction than on approach because it takes a finite time for the deformed cell surfaces to relax and to recover to their original shape. According to the modeling of the force–displacement process, as described in the previous section, the deformation–relaxation process can be plotted as shown in Figure 5.8. At the given deformation $x_m = 170$ nm, an increase of the viscous factor η (from 0.35 to 0.45) will result in increased relaxation time. In other words, it takes longer for a cell with higher viscosity to recover from deformation, meaning at any given displacement, there will be a smaller repulsive force, resulting in a larger hysteresivity (ζ) and larger energy dissipation ($A - R$). This means that the cellular body will display increased viscous character when it has increased energy dissipation after EGF stimulation.

The energy dissipation from the QCM-D measurement is the sum of the energy dissipated in disruption of cellular adhesion E_a and the energy loss in the viscoelastic cytoplasm E_c: $E = E_a + E_c$. The EGF reduces the focal adhesion area and strength in A431 cells [3,29], which results in the decrease in E_a as shown by the adhesion peeling model described in previous section.

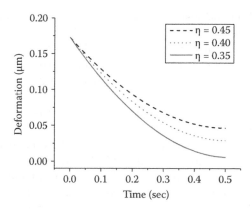

FIGURE 5.8
AFM energy dissipation simulation result shows that it takes longer time for the deformation to recover for a cell with higher viscosity η as indicated with η = 0.45 showing larger deformation overall.

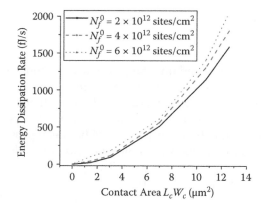

FIGURE 5.9
Cell peeling model simulation.

The simulation result in Figure 5.9 shows that the decrease in contact area A (x axis) as well as the reduced substrate adhesion ligand density (from 6×10^{12} to 2×10^{12} site/cm^2) will reduce the energy dissipation rate $\dot{E}_a$ (y axis). On the other hand, the loss of focal adhesion will cause cell rounding and swelling [3]. Our AFM nanoindentation measurements revealed that the cell body becomes more viscous after EGF stimulation. The increase in viscosity, η, in cell cytoplasm would induce relatively high damping and would increase the cytoplasm energy loss during the QCM-D measurement [15]. However, the QCM-D indicated a decrease in energy dissipation after EGF treatment, indicating that E_a dominates the energy dissipation process and E_c has minimal effect. The modeling of the wave propagation through the cellular material therefore should employ the multilayer configuration

(Figure 5.4c). Cellular adhesion through focal adhesion complexes linking the cytoskeleton to the extracellular matrix (ECM) creates an interfacial layer between the cell membrane and the ECM, with a gap thickness ranging from 30 nm in the closest contact area to 100–250 nm in the remaining areas [30–32].

Evidence shows that the acoustic wave decays almost completely before reaching the cell cytoplasm [32], allowing energy dissipation in cellular adhesion to become the primary energy loss mechanism during the QCM-D measurement, which agrees with our experimental findings. Therefore, the characterization process benefits from the difference in measurement mechanisms of the two instruments. AFM performs nanoindentation on cell membranes with less than 500 nm deformation in a single-cell analysis fashion revealing cell cytoplasm properties, while QCM-D monitors the cell–ECM interfacial layer where focal adhesion dominates the energy dissipation. Therefore, AFM and QCM-D work in a complementary manner in the characterization of the changes in viscoelastic properties of the cell.

5.4 Conclusion

Cell signaling is one of the fundamental processes that control cell fate. It regulates the cell shape, and in turn, the cell mechanics. To identify the dynamic signaling pathway in situ, instruments and techniques that are capable of monitoring real-time mechanical property changes (such as viscoelasticity), as well as structural rearrangements, are crucial. AFM has been proven to be an effective instrument for visualizing cell membrane and cytoskeleton structures. It can also provide mechanical information with high temporal and spatial resolution. Meanwhile, the changes in the viscoelasticity can also be measured as the energy dissipation change of a cell monolayer by means of QCM-D. Our study pioneered a novel approach where AFM and QCM-D work in complementary fashion to characterize cellular viscoelastic properties during signaling-induced cell cytoskeleton remodeling. The use of these two techniques together facilitates the investigation of structural and mechanical biomarkers in cellular signaling processes. Knowledge of the changes in mechanical properties will provide us with more insight into the dynamics of cell signaling.

Acknowledgments

This research work was partially supported under NSF Grants IIS-0713346. The project described was also supported by Grant Number R43 GM084520 from the National Institute of General Medical Sciences of the NIH.

The authors would like to thank Dr. Chanmin Su of Veeco Instrument Inc. for his technical advice and help during the process of this research.

References

1. P. S. Grudinkin, V. V. Zenin, A. V. Kropotov, V. N. Dorosh, and N. N. Nikolsky, "Egf-induced apoptosis in a431 cells is dependent on stat1, but not on stat3," *European Journal of Cell Biology*, vol. 86, no. 10, pp. 591–603, October 2007.
2. K. Oda, Y. Matsuoka, A. Funahashi, and H. Kitano, "A comprehensive pathway map of epidermal growth factor receptor signaling," *Molecular Systems Biology*, vol. 1, 2005, doi: 10.1038/msb4100014384003.
3. M. Chinkers, J. Mckanna, and S. Cohen, " Rapid rounding of human epidermoid carcinoma-cells A-431 induced by epidermal growth-factor," *Journal of Cell Biology*, vol. 88, no. 2, pp. 422–429, 1981.
4. P. Janmey, "The cytoskeleton and cell signaling: Component localization and mechanical coupling," *Physiological Reviews*, vol. 78, no. 3, pp. 763–781, July 1998.
5. S. E. Cross, Y.-S. Jin, J. Rao, and J. K. Gimzewski, "Nanomechanical analysis of cells from cancer patients," *Nature Nanotechnology*, vol. 2, no. 12, pp. 780–783, December 2007.
6. A. Ridley, M. Schwartz, K. Burridge, R. Firtel, M. Ginsberg, G. Borisy, J. Parsons, and A. Horwitz, "Cell migration: Integrating signals from front to back," *Science*, vol. 302, no. 5651, pp. 1704–1709, December 5, 2003.
7. D. Ingber, "Tensegrity I. Cell structure and hierarchical systems biology," *Journal of Cell Science*, vol. 116, no. 7, pp. 1157–1173, April 1, 2003.
8. N. Wang, K. Naruse, D. Stamenovic, J. Fredberg, S. Mijailovich, I. Toric-Norrelykke, T. Polte, R. Mannix, and D. Ingber, "Mechanical behavior in living cells consistent with the tensegrity model," *Proceedings of the National Academy of Sciences of the United States of America*, vol. 98, no. 14, pp. 7765–7770, July 3, 2001.
9. G. Binnig, C. Quate, and C. Gerber, "Atomic force microscope," *Physical Review Letters*, vol. 56, no. 9, pp. 930–933, March 3, 1986.
10. E. Dimitriadis, F. Horkay, J. Maresca, B. Kachar, and R. Chadwick, "Determination of elastic moduli of thin layers of soft material using the atomic force microscope," *Biophysical Journal*, vol. 82, no. 5, pp. 2798–2810, May 2002.
11. C. K. M. Fung, K. Seiffert-Sinha, K. W. C. Lai, R. Yang, D. Panyard, J. Zhang, N. Xi, and A. A. Sinha, "Investigation of human keratinocyte cell adhesion using atomic force microscopy," *Nanomedicine-Nanotechnology Biology and Medicine*, vol. 6, no. 1, pp. 191–200, February 2010.
12. D. J. Mueller and Y. F. Dufrene, "Atomic force microscopy as a multifunctional molecular toolbox in nanobiotechnology," *Nature Nanotechnology*, vol. 3, no. 5, pp. 261–269, May 2008.
13. S. E. Cross, Y.-S. Jin, J. Tondre, R. Wong, J. Rao, and J. K. Gimzewski, "AFM-based analysis of human metastatic cancer cells," *Nanotechnology*, vol. 19, no. 38, September 24, 2008. doi: 10.1058/09574484/19/38/384003.
14. E. Barthel, "Adhesive elastic contacts: JKR and more," *Journal of Physics D-Applied Physics*, vol. 41, no. 16, August 21, 2008.

15. F. Hook and B. Kasemo, "The QCM-D technique for probing biomacromolecular recognition reactions," in *Piezoelectric Sensors*, Steinem, C. and Janshoff, A., Eds., 2007, vol. 5, Springer, New York, pp. 425–447.

16. M. Edvardsson, M. Rodahl, and F. Hook, "Investigation of binding event perturbations caused by elevated QCM-D oscillation amplitude," *Analyst*, vol. 131, no. 7, pp. 822–828, 2006.

17. K. Glasmastar, C. Larsson, F. Hook, and B. Kasemo, "Protein adsorption on supported phospholipid bilayers," *Journal of Colloid And Interface Science*, vol. 246, no. 1, pp. 40–47, February 1 2002.

18. C. Marxer, M. Coen, T. Greber, U. Greber, and L. Schlapbach, "Cell spreading on quartz crystal microbalance elicits positive frequency shifts indicative of viscosity changes," *Analytical and Bioanalytical Chemistry*, vol. 377, no. 3, pp. 578–586, October 2003.

19. M. Voinova, M. Rodahl, M. Jonson, and B. Kasemo, "Viscoelastic acoustic response of layered polymer films at fluid-solid interfaces: Continuum mechanics approach," *Physica Scripta*, vol. 59, no. 5, pp. 391–396, May 1999.

20. V. Saravia and J. L. Toca-Herrera, "Substrate influence on cell shape and cell mechanics: HepG2 cells spread on positively charged surfaces," *Microscopy Research and Technique*, vol. 72, no. 12, pp. 957–964, December 2009.

21. O. Hayden, R. Bindeus, and F. Dickert, "Combining atomic force microscope and quartz crystal microbalance studies for cell detection," *Measurement Science and Technology*, vol. 14, no. 11, pp. 1876–1881, November 2003.

22. S. Notley, M. Eriksson, and L. Wagberg, "Visco-elastic and adhesive properties of adsorbed polyelectrolyte multilayers determined in situ with QCM-D and AFM measurements," *Journal of Colloid and Interface Science*, vol. 292, no. 1, pp. 29–37, December 1, 2005.

23. P. Hansma, H. Yu, D. Schultz, A. Rodriguez, E. A. Yurtsev, J. Orr, S. Tang, J. Miller, J. Wallace, F. Zok, C. Li, R. Souza, A. Proctor, D. Brimer, X. Nogues-Solan, L. Mellbovsky, M. J. Pena, O. Diez-Ferrer, P. Mathews, C. Randall, A. Kuo, C. Chen, M. Peters, D. Kohn, J. Buckley, X. Li, L. Pruitt, A. Diez-Perez, T. Alliston, V. Weaver, and J. Lotz, "The tissue diagnostic instrument," *Review of Scientific Instruments*, vol. 80, no. 5, May 2009.

24. O. Klymenko, J. Wiltowska-Zuber, M. Lekka, and W. M. Kwiatek, "Energy Dissipation in the AFM Elasticity Measurements," *Acta Physica Polonica A*, vol. 115, no. 2, pp. 548–551, February 2009, 42nd Zakopane School of Physics International Symposium Breaking Frontiers, Zakopane, Poland, May 19–25, 2008.

25. D. Johannsmann, "Viscoelastic, mechanical, and dielectric measurements on complex samples with the quartz crystal microbalance," *Physical Chemistry Chemical Physics*, vol. 10, no. 31, pp. 4516–4534, 2008.

26. C. Dong and X. Lei, "Biomechanics of cell rolling: Shear flow, cell-surface adhesion, and cell deformability," *Journal of Biomechanics*, vol. 33, no. 1, pp. 35–43, January 2000.

27. M. Dembo, D. Torney, K. Saxman, and D. Hammer, "The reaction-limited kinetics of membrane-to-surface adhesion and detachment," *Proceedings of the Royal Society B-Biological Sciences*, vol. 234, no. 1274, pp. 55–83, June 22, 1988.

28. M. Puig-De-Morales-Marinkovic, K. T. Turner, J. P. Butler, J. J. Fredberg, and S. Suresh, "Viscoelasticity of the human red blood cell," *American Journal of Physiology-Cell Physiology*, vol. 293, no. 2, pp. C597–C605, August 2007.

29. H. Xie, M. Pallero, K. Gupta, P. Chang, M. Ware, W. Witke, D. Kwiatkowski, D. Lauffenburger, J. Murphy-Ullrich, and A. Wells, "EGF receptor regulation of cell motility: EGF induces disassembly of focal adhesions independently of the motility-associated PLC gamma signaling pathway," *Journal of Cell Science*, vol. 111, no. Part 5, pp. 615–624, March 1998.

30. K. Giebel, C. Bechinger, S. Herminghaus, M. Riedel, P. Leiderer, U. Weiland, and M. Bastmeyer, "Imaging of cell/substrate contacts of living cells with surface plasmon resonance microscopy," *Biophysical Journal*, vol. 76, no. 1, Part 1, pp. 509–516, January 1999.

31. F. Li, J. H. C. Wang, and Q.-M. Wang, "Thickness shear mode acoustic wave sensors for characterizing the viscoelastic properties of cell monolayer," *Sensors and Actuators B-Chemical*, vol. 128, no. 2, pp. 399–406, January 15, 2008.

32. J. Wegener, J. Seebach, A. Janshoff, and H. Galla, "Analysis of the composite response of shear wave resonators to the attachment of mammalian cells," *Biophysical Journal*, vol. 78, no. 6, pp. 2821–2833, June 2000.

6

Modeling and Experimental Verifications of Cell Tensegrity

Ning Xi, Ruiguo Yang, Carmen Kar Man Fung, King Wai Chiu
Lai, Bo Song, Kristina Seiffert-Sinha, and Animesh A. Sinha

CONTENTS

6.1 Introduction

The tensegrity cell model is a qualitative modeling technique for adherent cells [1,2]. It has been proven to be consistent with the mechanical behavior of living cells [3]. The main concept in delineating the mechanical properties of cells is that the cell body is a tensional integrity structure, thus the name *tensegrity* [4]. The cytoskeleton behaves as a discrete mechanical network composed of three elements: microtubule, actin filament, and intermediate filament [5]. The three elements have different mechanical properties. The microtubule is the stiffest of the three and thus bears the compressional force when the cytoskeleton is compressed, while the actin filament and intermediate filament assume more tension when the cell is stretched [4]. Through the interconnected cytoskeletal network, cells maintain the structure integrity under different force interactions.

It is generally accepted that a cell can establish tensegrity force balance with its cytoskeletal network and the extracellular matrix (ECM) [1]. The

ECM provides the external support for the force balance, in which the integrin of cells binds to ECM in the formation of the so-called focal adhesion complex [6]. The focal adhesion sites are discrete, nonuniform protrusion regions of the cell periphery [7]. They provide the linkage between the cell and ECM, and at the same time serve as the signaling transduction hub. They also host a cluster of proteins that behave as force sensors [7].

In tensegrity force balance, prestress in the filaments is a major factor governing cell mechanics. Under a certain force balance or structural configuration, a cell will have a corresponding prestress, and ultimately a certain level of elasticity that can generally be revealed by AFM probing. Thus, by measuring the elasticity of the cell either locally or globally [8], the prestress of the cell can be determined and its physical configuration can be predicted. Therefore, to determine whether cells can be modeled as tensegrity structures, one would need to tune the cell prestress by biochemically modifying its living conditions and monitoring whether the elasticity changes accordingly.

Metastatic adenocarcinoma cells show anchorage-independent growth patterns, and they become round in shape, whereas normal mesothelial cells show a large and more stretched morphology. Experiments have verified that the cancerous type is 70% less stiff [9]. From the force-balancing perspective, cells with weak focal adhesions should be less prestressed than cells with stronger and denser focal adhesions. An analogy can be drawn for the cells with focal adhesion of different strengths and a tent set up with a varied number of anchors to the ground. Another piece of evidence that signifies the difference in prestress lies in the cytoskeleton structure itself; when cells are less prestressed, they tend to have a disorganized actin filament network [10], a sign of less tensional forces in individual actin filaments.

The tensegrity modeling of adherent cells is normally about single cells, and only cell–ECM interaction is considered in most cases. Cell–cell interactions, although often ignored when considering cell anchorage, can also be a source of external support in terms of force balancing. As a special type of cell–cell adhesion structure, desmosome is one of the main cell–cell adhesion complexes in epithelial cells that provides mechanical strength to maintain the tissue integrity. Desmosome links the intermediate filaments of neighboring cells through a cadherin type of adhesion. From the development point of view, desmosomes from different layers of epidermis vary in size and appearance: the basal layer epidermis has less-organized desmosomes with smaller dimensions compared with desmosomes in the suprabasal layers [11]. Apparently, cells at the suprabasal layers are tougher and stronger. We speculate that when the desmosome becomes denser, cellular tension becomes stronger and the prestress becomes larger. All this evidence points to the idea that cell–cell adhesions, especially desmosomes, play an important role in providing external support just as focal adhesions do. In this study, we put this hypothesis to the test by modulating the cell–cell adhesions between neighboring cells using biochemical and mechanical methods, and were able to obtain quantitative stiffness data validating the change of cellular stiffness.

6.2 Decrease in Cell Stiffness Resulting from Desmosome Disruption

Keratinocyte is a major constituent of the epidermis and is abundant in desmosome-based cell–cell adhesions. As mentioned earlier, desmosomes provide the strength to maintain the mechanical integrity of the tissue by "spot-welding" the intermediate filaments from neighboring cells. Desmosomes are composed of a cluster of proteins to form a molecular structure. The desmosome complex employs cytoplasmic protein plaque to tether to the intermediate filament as illustrated in Figure 6.1. To disrupt the connection between neighboring cells would require the disassembly of the desmosome, which would release the intermediate filament linkage, which normally plays the role of anchor points. The disassembly of the desmosome would therefore remove the anchor points from the tensegrity structure and ultimately reduce the amount of stress in the system. This hypothesis was verified by two biochemical experiments.

6.2.1 Decrease in Stiffness Resulting from Desmosome Disassembly

Desmosomes sometimes can be damaged in certain diseases, one of which is pemphigus vulgaris (PV), a potentially fatal autoimmune disease that targets

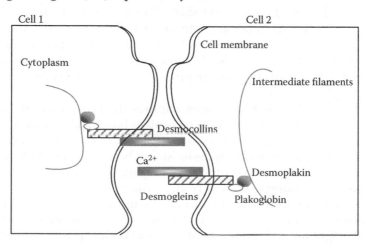

FIGURE 6.1
Desmosome structure with different desmosomal proteins connecting each other all the way to the intermediate filaments.

one of the proteins in the complex, Dsg3 [12]. The binding of the autoimmune antibody causes the release of the protein from the complex and eventually leads to the disassembly of the desmosome structure. The damaged desmosome leads to the loss of cellular adhesion between neighboring keratinocyte cells and eventually causes blistering of the skin from the body fluid underneath the cells.

We tested the effect of the autoimmune antibody in vitro on keratinocytes. The administration of anti-Dsg3 antibody induced the loss of cellular adhesion structures. This was confirmed by comparing images obtained before and after the administration of the antibody. The AFM images show the enlarged spacing between neighboring cells, clear evidence of cell adhesion loss, as well as retracted filament structures from the cell periphery [13]. The topographical change is accompanied by a decrease in the cell stiffness as measured by AFM-based nanoindentation. The Young's modulus value of normal cells dropped from 29.8 ± 4.9 to 18.6 ± 4.6 kPa for anti-Dsg3 antibody-treated cells.

In the AFM nanoindentation experiment, a silicon nitride cantilever (Bruker-nano, Santa Barbara, CA) was used, the spring constant of which is 0.06 N/m calibrated by the thermal tune method. The applied force can be calculated by Hooke's law as $F = kd$, where d is the deflection of the cantilever and k is the spring constant of the cantilever. According to the Hertzian model, for a cone-shaped tip, the relation between indentation and deformation is defined as [14]

$$F = \frac{2}{\pi} \frac{E}{1 - \upsilon^2} \delta^2 tan\alpha \qquad (6.1)$$

where α is the half angle of the cone-shaped tip, υ is the Poisson ratio, F is the applied force, δ is the indentation depth, and E is the Young's modulus value. Young's moduli can be generated by fitting the force–displacement curve. The half-open angle of the tip is $17.5°$ and we used 0.5 as the Poisson ratio. The force–displacement curves were processed with a MATLAB® routine to obtain Young's modulus.

Since the formation of desmosomes requires the presence in the growth medium of Ca^{2+}, which would form a covalent bond with the desmosomal protein Dsg [15] as shown in Figure 6.1, depletion of the Ca^{2+} in the growth medium hinders the formation of desmosomes between cells. AFM imaging shows a significant difference in terms of dimensions for cells cultured in normal growth medium and those from media with Ca^{2+} depleted. Normal cells are stretched with a diameter around 25 µm and a height around 1 µm, while calcium-depleted cells are rounded up with a diameter around 12 µm and a cell height around 3 µm. The stiffness of both types of cells was also measured and compared. The Young's modulus values agree well with previous measurements for normal cells and antibody-treated cells. The stiffness

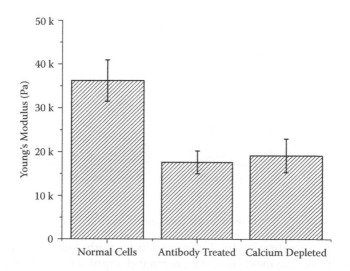

FIGURE 6.2
Young's modulus comparison for normal and calcium-depleted cells.

decreases from 35.6 ± 3.5 kPa to around 22.3 ± 2.9 kPa for normal and calcium-depleted cells in Figure 6.2. These biochemical methods indeed verified that disruption of the desmosome would lead to a decrease in cellular stiffness, an indication of reduced prestress in the tensegrity-based cytoskeletal structure. The release of the intermediate filament connection was induced by desmosome disassembly or nonformation. We then demonstrated that by severing the intermediate filament mechanically and directly we could also achieve a similar effect.

6.2.2 Decrease in Stiffness Resulting from AFM-Based Nanosurgery

Imaging of AFM is the direct result of the scanning motion driven by the XY direction piezo actuation unit. From the motion control perspective, the zigzag-shaped driving voltage is applied to the XY piezo scanner, causing a linear motion back and forth with a predefined frequency. However, this is not the case for a nanomanipulation operation, where the position of the tip is not linearly related to an applied signal/voltage and the applied voltage should be an arbitrary shape rather than a zigzag. In addition, most nanomanipulation operations on biological samples are performed in liquid. Thus, the viscoelasticity of the sample requires a higher response frequency, and the control system should be upgraded to meet this demand. The upgraded hardware and software configurations convert a commercial AFM to a nanobiomanipulation-oriented AFM. The technical details of the development of motion controllers can be found in [16]. In brief, we used a signal access module to provide all the signals needed for the external control of the

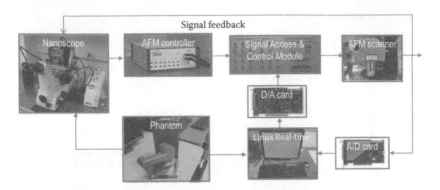

FIGURE 6.3
AFM-based nanorobotic system used for the nanodissection.

manipulation operation by the Linux controller with data acquisition cards as the interface. The joystick is the command input where all the motions start, enabling the precision manipulation of nanoscale objects with force feedback (diagram shown in Figure 6.3). By utilizing the AFM tip as a robotic arm, the capability of AFM can be extended greatly.

By human-controlled AFM nanorobotic operation, the intermediate filaments can be dissected bundle by bundle to isolated neighboring cells. Bundles of intermediate filaments can be cut off precisely with less than 100 nm resolution in height by controlling the AFM tip position over the intercellular junction area, and maintaining a constant normal force that is large enough to penetrate the cell membrane; the resulting lateral force would therefore sever the filament structures underlying the cell membrane. The intermediate filament bundles on the top were removed and replaced with a trench. The height difference is around 100 nm as magnified in the cross section with the circles. Intermediate filaments from individual cells, therefore, can be dissected entirely or partially by repeating the nanosurgery operation.

Stiffness data was also collected before and after the dissection operation from AFM-based nanoindenation. The obtained Young's modulus value decreases from 32.1 ± 2.8 kPa to 18.8 ± 2.0 kPa (Figure 6.4). The well-connected cells with intact cell–cell adhesion are around 1.5 times stiffer than the cells whose intermediate filaments are cut loose. When cells are modeled as tensegrity, prestress determines the mechanical response of the whole cytoskeleton structure to external load such as the nanoindentation force from the probe. With less external support from intermediate filaments of their neighbors, cells would have to establish a new force balance, one which presumably receives no contribution from tension in the intermediate filaments. Theoretical calculation was then performed to verify the stiffness change with and without intermediate filament tensions in the structural model.

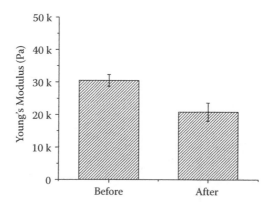

FIGURE 6.4
The stiffness response before and after AFM-based nanodissection.

6.3 Quantitative Modeling Based on Six Struts Tensegrity Structure

The structure model shown in Figure 6.5 with 6 struts (thick lines) and 24 cables (thin lines) was originally developed by Ingber [17] as a conceptual model for cellular tensegrity (Figure 6.5a). It is constructed by connecting the ends of compressional struts with the tensional cables. The struts mimic the role of the microtubule in the cytoskeleton, while the cables denote the actin filaments. Considering that actin filaments normally lie beneath the cell membrane and are abundant at the cell peripheral [5], the design ensures that all the tensional cables for microfilaments are at the boundary of the structure. Since in each direction we have a symmetric structure of two struts, the same notation is used for both struts in each direction, as AA, BB, and CC. The length of each strut is L, and the distance between them is $s = L/2$. The distance between each pair of notes (the cable length of AB, BC, or AC) is l.

To quantify the relationship between the stress and the strain, a signature characteristic of mechanical behavior, parallel loads are applied to both ends of the structure in one direction. A force of $T/2$ is applied to each node of C along the x direction to stretch the structure. At this configuration and with the applied load, we want to find out the relationship between the strain and the stress, or in other words, the modulus in the load direction.

By the virtual work method, we define that the work of T on an incremental extension δ_{s_x} per unit reference volume V of the model equals the work of uniaxial stress on the incremental change in uniaxial strain δ_{e_x}, a relationship captured in Equation 6.2.

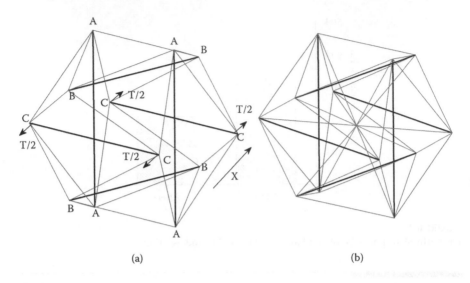

(a)   (b)

FIGURE 6.5
(a) The configuration of the tensegrity structure without intermediate filaments, the thick lines
indicate the microtubule as the compressive elements and the thin lines indicate the actin filament as the tensional elements; (b) the structure with intermediate filament (originated from
the center).

$$\frac{T\delta_{sx}}{V} = \sigma_X \delta_{ex} \tag{6.2}$$

where e_X and s_X denote the strain and extension in the X direction, and σ_X
denotes stress in the same direction. Since we know the dimensions of the
structure, we can calculate the volume (V) enclosed by it as in Equation 6.3:

$$V = \frac{5L^3}{16} \tag{6.3}$$

Strain is defined as the unit extension, so we have

$$\delta_{ex} = \frac{\delta_{sx}}{s} \tag{6.4}$$

From Equation 6.4 and considering the distance between the parallel struts,
it follows that

$$\delta_{ex} = 2\frac{\delta_{sx}}{L}$$

Through substitution, Equation 6.3 becomes

$$\frac{16T\delta_{sx}}{5L^3} = \sigma_X \frac{2\delta_{sx}}{L} \tag{6.5}$$

Thus,

$$\sigma_X = \frac{8T}{5L^2} \tag{6.6}$$

By definition of the modulus as the ratio between strain and stress in the X direction:

$$E^* = \frac{d\sigma_X}{de_X} = s\frac{d\sigma_X}{ds_X} = \frac{L}{2}\frac{d\sigma_X}{ds_X} \tag{6.7}$$

Taking the derivative of Equation 6.6, we obtain the relationship between strain and stress:

$$\frac{d\sigma_X}{ds_X} = \frac{8}{5L^2}\frac{dT}{ds_X} \tag{6.8}$$

Substituting Equation 6.8 into Equation 6.7, we arrive at the expression of the modulus as

$$E^* = \frac{L}{2}\frac{8}{5L^2}\frac{dT}{ds_X} = \frac{0.8}{L}\frac{dT}{ds_X} \tag{6.9}$$

The E^* in Equation 6.9 is the Young's modulus obtained from the structural derivation, and it will be applied to the structure both with and without intermediate filaments by plugging in the dT/ds_X from the two structures.

6.3.1 Without Intermediate Filaments

From Equation 6.9, we found the expression of E^* that is written in terms of dT/ds_X. We only need to find the relation between the force T and the distance between the struts s_X. For the configuration without the intermediate filaments, the calculation is as follows. The relationship between the force T and the extension is

$$T = 2F_{AB}\frac{s_X - L_{BB}}{l_{AB}} + 2F_{AC}\frac{s_X}{l_{AC}} \tag{6.10}$$

where F_{AB} and F_{AC} are the forces in the corresponding cable. The kinematics of the configuration in Figure 6.5a is

$$l_{AB} = \frac{1}{2}\sqrt{(L_{BB} - s_X)^2 + (s_Y)^2 + (L_{AA})^2}$$

$$l_{AC} = \frac{1}{2}\sqrt{(L_{AA} - s_Z)^2 + (s_X)^2 + L_{CC}^2} \qquad (6.11)$$

$$l_{BC} = \frac{1}{2}\sqrt{(L_{CC} - s_Y)^2 + (s_Z)^2 + L_{BB}^2}$$

If we assume that the cables are linearly elastic of stiffness k and resting length l_r, the force then in each cable is given as

$$F = \begin{cases} k(l - l_r) & \text{if } l > l_r \\ 0 & \text{if } l < l_r \end{cases} \qquad (6.12)$$

By plugging the kinematics of the structure in Equation 6.11 and the force expression in Equation 6.12, we have the relationship between T and s_X. Taking the derivation of that, we can obtain dT/ds_X. This derivation will be evaluated in the initial state (the reference state) where the parameters are

$$l_{AB} = l_{AC} = l_{BC} = l_0$$

$$L_{AA} = L_{BB} = L_{CC} = L_0$$

$$s_X = s_Y = s_Z = \frac{L_0}{2} \qquad (6.13)$$

$$l_0 = \sqrt{\frac{8}{3}} L_0$$

After plugging in these parameters, we obtain

$$\left. \frac{dT}{ds_X} \right|_0 .$$

Plugging into Equation 6.9 we have the definition of E^* in the configuration without the intermediate filament

$$E^* \doteq 15.6 \frac{F}{L_0^2} \frac{1 + 4\varepsilon_0}{1 + 12\varepsilon_0} \qquad (6.14)$$

where ε_0 is the initial cable strain, defined as

$$\varepsilon_0 = \frac{l - l_r}{l}$$

6.3.2 With Intermediate Filaments

The structures with intermediate filaments are shown in Figure 6.5b. The lines connecting the nodes in the middle are intermediate filaments. This structure is introduced by considering the fact in biology that the intermediate filaments are projecting from the nucleus to the periphery of the cell [5]. Thus, the designed structure has a tensional cable connecting each pair of struts and the cable all crossing the center O. With these additional structural components, the kinematics for the actin filament remains, while the kinematics for the intermediate filament can be characterized as

$$r_{OA} = \frac{1}{2}\sqrt{s_x^2 + L_{AA}^2}$$

$$r_{OB} = \frac{1}{2}\sqrt{s_y^2 + L_{BB}^2} \tag{6.15}$$

$$r_{OC} = \frac{1}{2}\sqrt{s_z^2 + L_{CC}^2}$$

With the addition of intermediate filament structures, some force balance establishment has to take into consideration the tension in the intermediate filament forces in N_{OA}, N_{OB}, and N_{OC}. The force balance now has been reconfigured and can be expressed as

$$T = 2\left(F_{AB}\frac{s_x - L_{BB}}{l_{AB}} + F_{AC}\frac{s_x}{l_{AC}} + N_{OA}\frac{s_x}{r_{OA}} \right)$$

$$F_{BC}\frac{L_{BB} - s_y}{l_{BC}} - F_{AB}\frac{s_y}{l_{AB}} - N_{OB}\frac{s_y}{r_{OB}} = 0 \tag{6.16}$$

$$F_{AC}\frac{L_{AA} - s_z}{l_{AC}} - F_{BC}\frac{s_z}{l_{BC}} - N_{OC}\frac{s_z}{r_{OC}} = 0$$

The force in each of the tensional elements including the actin filaments and the intermediate filaments can then be expressed by Hooke's law. For actin filaments, it remains the same as the previous configuration in Equation 6.12. For intermediate filaments, a second-order elastic element was added with elastic constant k_2 on the basis of a first-order elastic element:

$$F = \begin{cases} k_1(r - r_r) + k_2(r - r_r)^2 & \text{if } r > r_r \\ 0 & \text{if } r < r_r \end{cases} \tag{6.17}$$

where r_r is the reference length (zero tensional force) for the intermediate filaments.

To find the relationship between the applied load T and the displacement s_x, we can plug the kinematics into the force balance in Equation 6.16 and replace the force with the expressions in Equation 6.17. We then have the static balance condition with the intrinsic parameter of the components and the kinematic. Using the same procedure as in the previous configuration, we can take the spatial derivative of T with respect to s_X to obtain dT/ds_X. This derivative will be evaluated at the initial state with the following parameters:

$$l_{AB} = l_{Ac} = l_{BC} = l_0$$
$$L_{AA} = L_{BB} = L_{CC} = L_0$$
$$s_X = s_Y = s_Z = L_0 / 2 \tag{6.18}$$
$$l_0 = \sqrt{3/8}L_0$$

After evaluation under the initial conditions, the derivative

$$\left.\frac{dT}{s_X}\right|_0$$

can be plugged into Equation 6.9, so that we have E^*_{with}:

$$E^* = \frac{L}{2}\frac{8}{5L^2}\frac{dT}{ds_x} = E_{F_0} + E_{F_1} + E_{F_2} \tag{6.19}$$

where

$$E_{F_0} = 0.33\frac{F_0}{L_0^2}\frac{6\varepsilon_0 + 1}{\varepsilon_0}$$

defines the contribution from the stress in the actin filament;

$$E_{F_1} = 0.01\frac{F_1}{L_0^2}\frac{43\varepsilon_0 + 2}{\varepsilon_0}$$

defines the contribution from the linear portion (F_1, k_1) of the intermediate filament; and

$$E_{F_2} = \frac{F_2}{L_0}\frac{1.6\varepsilon_0 + 0.3}{\varepsilon_0(\varepsilon_0 + 1)}$$

defines the contribution from the quadratic portion (F_2, k_2) of the intermediate filament. Compared with E_{F_0} and E_{F_1}, E_{F_2} is much smaller and therefore can be neglected. We can assume that the intermediate filament can also be modeled as a first-order Hooke relation. However, E_{F_1} and E_{F_0} are comparable, and thus E_{F_1} will make a difference for the modulus with an intermediate filament structure.

TABLE 6.1

Parameter List for the Numerical Evaluation

Variable	Physical Meaning	Value	Unit
F_0	Average force generated by a single actomyosin motor	1.6	nN
F_1	Average force generated by a single intermediate filament from linear portion	3.2	nN
F_1	Average force generated by a single intermediate filament from quadratic portion	0.8	nN
L_0	Estimated length of an actin filament from volume fraction	0.7	μm
K	The stiffness coefficient of an actin filament	0.05	pN/nm
k_1	The linear portion of stiffness coefficient of an intermediate filament	0.10	pN/nm
k_2	The quadratic portion of stiffness coefficient of an intermediate filament	0.25	pN/nm

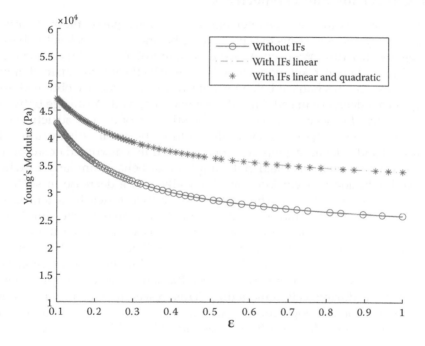

FIGURE 6.6

Young's modulus comparison with and without intermediate filaments from simulation based on the structure model.

With the intrinsic parameters of individual filaments obtained from [17], we can calculate the Young's modulus values numerically. The parameters used in the calculation are shown in Table 6.1. The plotting in Figure 6.6 shows the comparison of Young's modulus values with respect to the strain for three different situations, without intermediate filaments, with intermediate filaments in only the linear portion, and with intermediate filaments containing both linear and quadratic terms. The curve for the linear portion of the intermediate filaments overlaps with the one containing both linear and quadratic portions. This also confirms that E_{F_2} is negligible. The Young's modulus with intermediate filaments in the configuration increased by 18% at the upper boundary when $\varepsilon_0 \to 1$. The trend is also indicative of the contribution from the intermediate filaments; that is, when the strain increases, the stiffness difference also increases. This agrees well with the functional contribution of intermediate filaments in the cytoskeleton system. They are responsible for maintaining the shape when there is large strain or deformation [5].

6.4 Conclusion and Perspectives

In this study, we validated that intercellular adhesion is part of the mechanism employed by epithelial cells to achieve force balance in a cytoskeleton-based tensegrity structure. We used AFM-based nanorobotics to dissect the cell–cell adhesion between kerationcyte cells, primarily the intermediate filament structures. The mechanical behavior data were subsequently obtained and a significant decrease in cellular stiffness was observed. A linkage between the integrity of cell–cell adhesion and the stiffness of the cell structure can be built. The cell adhesion loss was also induced biochemically by antibody treatment and calcium depletion, which resulted in a disrupted desmosome structure. Both methods yield cell samples that behave similarly to when cell–cell adhesion was cut loose mechanically, and a decrease in stiffness was observed as expected. The findings were then tested by a tensegrity structural model with and without intermediate filaments; the simulation result verifies the reduction in Young's modulus values when there is no extra tensional element provided by the intermediate filament. Thus far, the hypothesis was validated that epithelial cells employ cell–cell adhesion as a complementary mechanism to cell–ECM adhesion to sustain their mechanical integrity during cell remodeling. This research may provide some insights into the development of new cell growth environments by taking into consideration cell–cell adhesion beyond cell–ECM interactions.

Acknowledgments

This research work was partially supported under NSF Grants IIS-0713346. The project described was also supported by Grant Number R43 GM084520 from the National Institute of General Medical Sciences of the NIH.

The authors would like to thank Dr. Chanmin Su of Veeco Instrument, Inc., for his technical advice and help during the course of this research.

References

1. D. Ingber, "Tensegrity I. Cell structure and hierarchical systems biology," *Journal of Cell Science*, vol. 116, no. 7, pp. 1157–1173, April 1, 2003.
2. N. Wang, K. Naruse, D. Stamenovic, J. Fredberg, S. Mijailovich, I. Toric-Norrelykke, T. Polte, R. Mannix, and D. Ingber, "Mechanical behavior in living cells consistent with the tensegrity model," *Proceedings of the National Academy of Sciences of the United States of America*, vol. 98, no. 14, pp. 7765–7770, July 3, 2001.
3. D. Ingber, "Tensegrity II. How structural networks influence cellular information processing networks," *Journal of Cell Science*, vol. 116, no. 8, pp. 1397–1408, April 15, 2003.
4. D. Ingber, "Tensegrity: The architectural basis of cellular mechanotransduction," *Annual Review of Physiology*, vol. 59, pp. 575–599, 1997.
5. B. Alberts, A. Johnson, J. Lewis, M. Raff, K. Roberts, and P. Walter, *Molecular Biology of the Cell*, Fifth Ed. New York: Garland Science, 2007.
6. G. Plopper, H. Mcnamee, L. Dike, K. Bojanowski, and D. Ingber, "Convergence of integrin and growth-factor receptor signaling pathways within the focal adhesion complex," *Molecular Biology of the Cell*, vol. 6, no. 10, pp. 1349–1365, October 1995.
7. M. Wozniak, K. Modzelewska, L. Kwong, and P. Keely, "Focal adhesion regulation of cell behavior," *Biochimica et Biophysica Acta-Molecular Cell Research*, vol. 1692, no. 2–3, SI, pp. 103–119, July 5, 2004.
8. P. Carl and H. Schillers, "Elasticity measurement of living cells with an atomic force microscope: Data acquisition and processing," *Pflugers Archiv-European Journal of Physiology*, vol. 457, no. 2, pp. 551–559, November 2008.
9. S. E. Cross, Y.-S. Jin, J. Rao, and J. K. Gimzewski, "Nanomechanical analysis of cells from cancer patients," *Nature Nanotechnology*, vol. 2, no. 12, pp. 780–783, December 2007.
10. F. Li, J. H. C. Wang, and Q.-M. Wang, "Thickness shear mode acoustic wave sensors for characterizing the viscoelastic properties of cell monolayer," *Sensors and Actuators B Chemical*, vol. 128, no. 2, pp. 399–406, January 15, 2008.
11. K. Green and C. Gaudry, "Are desmosomes more than tethers for intermediate filaments?" *Nature Reviews Molecular Cell Biology*, vol. 1, no. 3, pp. 208–216, December 2000.
12. C. Pincelli, "Apoptosis in pemphigus," *Journal of Investigative Dermatology*, vol. 126, no. 10, p. 2351, Oct. 2006.
13. C. K. M. Fung, K. Seiffert-Sinha, K. W. C. Lai, R. Yang, D. Panyard, J. Zhang, N. Xi, and A. A. Sinha, "Investigation of human keratinocyte cell adhesion using atomic force microscopy," *Nanomedicine-Nanotechnology Biology and Medicine*, vol. 6, no. 1, pp. 191–200, February 2010.

14. A. Touhami, B. Nysten, and Y. Dufrene, "Nanoscale mapping of the elasticity of microbial cells by atomic force microscopy," *Langmuir*, vol. 19, no. 11, pp. 4539–4543, May 27, 2003.
15. R. Windoffer, M. Borchert-Stuhltrager, and R. Leube, "Desmosomes: Interconnected calcium-dependent structures of remarkable stability with significant integral membrane protein turnover," *Journal of Cell Science*, vol. 115, no. 8, pp. 1717–1732, April 15, 2002.
16. R. Yang, N. Xi, K. Lai, B. Gao, H. Chen, C. Su, and J. Shi, "Motion controller for the Atomic Force Microscopy based nanomanipulation system," *IEEE International Conference on Intelligent Robots and Systems*, pp. 1339–1344, September 2009.
17. D. Stamenovic and M. Coughlin, "The role of prestress and architecture of the cytoskeleton and deformability of cytoskeletal filaments in mechanics of adherent cells: A quantitative analysis," *Journal of Theoretical Biology*, vol. 201, no. 1, pp. 63–74, November 7, 1999.

7

Modeling Swimming Micro/Nano-Systems in Low Reynolds Number

Stefan Nwandu-Vincent, Scott Lenaghan, and Mingjun Zhang

CONTENTS

7.1 Introduction

When the concept of swimming is presented, we naturally envision a high Reynolds number environment, mostly because our environment is that of a high Reynolds number. In this environment, inertial forces dominate and must be overcome in order to produce movement. Bacteria and other micro-organisms live in a low Reynolds number environment and are subject to different physical laws.

The Reynolds number is defined as

$$Re = \frac{\rho UL}{\eta} \tag{7.1}$$

where U = the mean velocity of the object relative to the fluid, L = characteristic linear dimension, η = viscosity of the fluid, and ρ = density of the fluid. When Equation (7.1) is multiplied by vL/vL, it becomes [1]

$$\text{Re} = \frac{\rho v L}{\eta} = \frac{\rho v^2 L^2}{\eta v L} = (\text{Inertia forces})/(\text{Viscous forces}) \qquad (7.2)$$

In a low Reynolds number environment, the viscous forces dominate the inertial forces. This effect can be seen in Equation 7.2, and the fluid passively responds to external forces. This means that a nonzero force on an object results in infinite acceleration [2]. As a result, the total net force and torque of an object in a low Reynolds number is always zero. This means that the movement of the object at a given moment is totally determined by the force exerted on it at that moment, and by nothing in the past [3]. Reciprocal motion in this type of environment does not lead to any progress. For an incompressible fluid the Navier–Stokes equations govern fluid movements,

$$\rho\left(\frac{\partial}{\partial t} + u \cdot \nabla\right) u = -\nabla p + \eta \nabla^2 u, \quad \nabla \cdot u = 0 \qquad (7.3)$$

with u being the flow field of the fluid and p the pressure of the surrounding medium. In low Reynolds number situations, the equation can be simplified to the Stokes equation,

$$-\nabla p + \mu \nabla^2 v = 0, \quad \nabla \cdot u = 0 \qquad (7.4)$$

Speed does not impact the motion because Equation 7.4 is time independent, so only body movements have an effect on the motion. If the body motion involves reciprocal shapes, which makes the process time reversible, no advancement in position occurs [2].

Bacteria and other micro-swimmers have developed numerous strategies to maneuver and navigate through this sort of environment; for example, prokaryotic cells swim using flagella motility, gliding, and twitching motility, while eukaryotic bacteria swim using flagella, cilia, or pseudopodia.

7.2 Prokaryotic Cell Swimming Strategies

7.2.1 Prokaryotic Flagella

Prokaryotic flagella are filamentous helical protein structures used for swimming in aqueous environments. Swimming speeds vary greatly between species. *Escherichia coli* swim at a rate of 25–35 μm/s, while *Bdellovibrio bacteriovorus* can reach up to 160 µm/s [4]. The flagellum is made up of the basal body, filament, and hook. The basal body acts as a motor. The motor, a reversible one, converts energy into useful mechanical motion similar to the motor found in cars and other mechanical devices. The flagella filament is rotated by this motor thus allowing the cell to swim. The hook couples the filament and the basal body. The energy for rotation is received from the gradients of ions across the cytoplasmic membrane; these ions are either protons or sodium. The motor works similar to a turbine, driven by the flow of ions [5]. These motors are exceptionally fast and have been reported to have speeds of up to 100,000 rpm. The hook connects the basal body and the filament and acts as a universal joint [4]. Most prokaryotic flagella can rotate both counterclockwise and clockwise, which contributes to their ability to change direction during swimming.

7.2.2 Twitching Motility

Twitching motility is a form of bacteria motility over moist surfaces that is independent of the use of the flagella. It occurs by extension, tethering, and then retraction of polar type IV pili, which operate in a manner similar to a grappling hook. It is mediated by type IV pili located at one or both poles of the cell and differs from flagella motility, which is mediated by the rotation of unipolar flagella [6]. Twitching motility is controlled by an array of signal transduction systems, including two-component sensor-regulators and a chemosensory system [6].

This type of motion was coined "twitching" because when it was first discovered in 1961, the cells appeared to move in a jerky fashion that resembled twitching [7].

Twitching motility is used by a wide range of bacteria, the best studied of which are *Pseudomonas aeruginosa, Neisseria gonorrhoeae,* and *Myxococcus xanthus,* where it is referred to as "social gliding motility" [6].

Twitching motility is used predominantly as a means for bacteria to travel in a low water content environment and to colonize hydrated surfaces, as opposed to free living in fluids [6]. It is vital for colonial behaviors such as the formation of biofilms and fruiting bodies [8].

7.2.3 Gliding Motility

Gliding motility is defined as a smooth translocation of cells over a surface by an active process that requires the expenditure of energy. It is flagella independent and follows the long axis of the cell [9]. Gliding bacteria live in environments as diverse as the human mouth, ocean sediments, and garden soil.

Studies have shown that gliding motility cannot be explained by a single mechanism. It appears that several different types of motors have evolved that allow movement over surfaces; some bacteria use type IV pilus extension and retraction, possibly powered by ATP hydrolysis, to propel themselves [9]. This process is sometimes put into the category of twitching motility, and it is a matter of debate if this system of motion should be considered twitching or gliding. Twitching is normally used to describe discontinuous cell movements, while gliding is used to describe smooth, continuous movements. A source for further confusion is that an organism that appears to twitch under certain conditions might appear to glide under other conditions [9]. Other forms of gliding that do not use pili exist. One example is mycoplasmas that depend on their well-developed cytoskeleton and surface adhesion proteins to crawl over surfaces. Some filamentous cyanobacteria's gliding motility may be powered by polysaccharide extrusion [9].

7.3 Eukaryotic Cell Swimming Strategies

7.3.1 Eukaryotic Flagella

Eukaryotic flagella are whiplike appendages that have a "9 + 2" structure, which is referred to as the *axoneme*; it has nine pairs of microtubule doublets surrounding two central single microtubules. At the base, it has a basal body that is the microtubule organizing center. The outer nine doublet microtubules extend a pair of dynein arms to the neighboring microtubule. The dynein arms cause flagella beating; through ATP hydrolysis, a force is produced by the dynein arms and results in the microtubule doublets sliding against each other and the whole flagellum bending. This mechanism is unlike the prokaryotic flagella working by a motor rotating the flagella. The eukaryotic flagella motion is usually planar and in the form of a traveling sinusoidal wave.

As an example of how the swimming of eukaryotic flagellum can be modeled, we will present an overview of the local resistive force theory and use it to understand Gray and Hancock's work on the propulsion of the sea urchin spermatozoa [10]. In order to model swimming in low Reynolds number situations, Equation 7.4, Stokes' equation, must be solved. Using Green's function for Stokes flow with a dirac-delta forcing of the form $\delta(x - x')F$, the fundamental solution to Stokes' equation is

$$u(x) = G(x - x') \cdot F \tag{7.5}$$

The Oseen tensor (G tensor) is

$$G(r) = \frac{1}{8\pi\eta}\left(\frac{1}{|r|} + \frac{|r||r|}{|r|^3}\right) \tag{7.6}$$

Equations (7.5) and (7.6) are referred to as a stokeslet, which represents a flow field due to F, a point force acting on the fluid at the position x' as a singularity [11]. To better understand the local drag theory we will use a representative model by Lauga and Powers [11]. The goal of the model is to find an approximate form for the resistance matrix of the flagella with errors controlled by the small parameter a/L, with a as the radius and L as the length [11]. The model assumes the flagellum to be a rod with length L and radius a subjected to an external force F_{ext} (uniformly distributed over the length of the rod); the rod is divided into equally spaced N points along the x-axis with the positions $x_j = (jL/N, 0, 0)$. These points are considered to be stokeslets, the far field flow induced by a short segment of the rod [11]. The velocity of each segment of the rod without taking into consideration the effects of other segments on it is $u = F_{ext}/\xi_{seg}$, where $\xi_{seg} \propto \eta a$ is the resistance coefficient of a segment. But since the flow induced by a segment affects all the other segments as well, the hydrodynamic interactions of the segments have to be included. The velocity of a segment u_j as a result of the jth segment is

$$u_j(x) = \frac{1}{8\pi\eta|x-x_j|}(1+e_xe_x)\cdot\left(\frac{F_{ext}}{N}\right) \tag{7.7}$$

So the velocity of a segment (ith segment) is a combination of the flows resulting from other segments and its velocity without the consideration of other segments,

$$u(x_i) = \frac{F_{ext}}{\xi_{seg}} + \sum_{j\neq i} u_j(x_i) \tag{7.8}$$

The sum over i is replaced by an integral taking into account that the density of the spheres is L/N and excluding a small region around x_i from the region of integration. This results in

$$u(x_i) = \frac{F_{ext}}{\xi_{seg}} + \frac{1}{8\pi\eta}\int' \frac{1}{|x_i-x|}(1+e_xe_x)\cdot F_{ext}\frac{dx}{L} \tag{7.9}$$

The prime above the integral signifies that the region of integration is between $-L/2$ to $L/2$ except for a region with size order a around x_i. With the end effects disregarded because $|x_i| \ll L$, the preceding Equation 7.9 becomes

$$u(x_i) = \frac{F_{ext}}{\xi_{seg}} + \frac{1}{4\pi\eta}\ln\left(\frac{L}{4a}\right)(1+e_xe_x)\cdot\left(\frac{F_{ext}}{L}\right) \tag{7.10}$$

Keeping only the terms that are leading orders in (L/a) and using the fact that $u(x_i)$ is constant for a rigid rod, the equation becomes

$$u = \frac{\ln(L/a)}{4\pi\eta}(1 + e_x e_x) \cdot \left(\frac{F_{ext}}{L}\right) \tag{7.11}$$

An assumption of the model is that there is no internal cohesive force acting between any pair of segments; only the hydrodynamic force is present. So the drag per unit length is $f = -F_{ext}/L$ and

$$f_N = -\xi_N u_N, \quad f_T = -\xi_T u_T$$
$$\xi_N = 2\xi_T = 4\pi\eta/\ln(L/a) \tag{7.12}$$

with N and T expressing the components normal and tangential to the x-axis. The movement of a thin filament in a viscous fluid elicits resistive forces in the normal and tangential directions of motion; an example of this phenomenon is illustrated in Figure 7.1.

The resistive force theory does have its limitations. Lauga and Powers [11] use the example of the sedimentation rate of a ring in a highly viscous fluid of radius R and rod diameter $2a$ (plane is horizontal) versus a straight rod of length $L = 2piR$ [11]. Even though the length of both rods is the same, the ring should fall faster because the segments of the ring are closer together, but using the drag coefficients of Equation 7.12 would have both falling at the same rate. This is because the above derivation assumes zero deformation in the filament. In filaments with a small curvature, it is assumed that the viscous force per unit length is the same as a rigid rod of the same length. So the

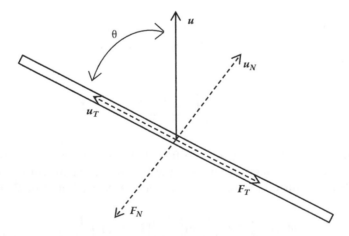

FIGURE 7.1
Representation of drag experienced by a slender body in a viscous fluid moving at a velocity u (see text for notation).

theory does not take into account the significant curvature of the ring. A way to improve this theory is to implement a combination of a smooth distribution of stokeslets and source dipoles to make a better approximation for the flow induced by the motion of the rod [11]. Gray and Hancock (henceforth referred to as G–H) used this approach to a sine wave with wavelength λ and derived the following equations for the resistive coefficients [10,12]

$$\xi_N = 2\xi_T = \frac{4\pi\eta}{n(2\lambda/a) - \frac{1}{2}} \tag{7.13}$$

G–H implemented this to analyze the propulsion of sea urchin spermatozoa. Their findings, recorded in the paper titled "The Propulsion of Sea Urchin Spermatozoa" [10], have since served as a basis for further work in flagellated microorganism propulsion studies.

The paper's objective was to investigate the forces the surrounding medium applied on the flagellum as the whole sperm swam and to relate the propulsive speed of the whole sperm to the form and speed of propagation of the bending waves generated by the tail. A spermatozoan's propulsion relies on the fact that the propulsive forces acting normally to the surface of the body offset the retarding tangential forces acting along the body. G–H use the forces elicited from the transverse displacement (V_y) impacted on a short element (δs) during the passage of a wave (Figure 7.2). V_y has two components: the tangential displacement $V_y \sin(\theta)$, and the displacement normal to the surface of the element $V_y \cos(\theta)$. The surrounding medium acts as a retarding force to both these displacements and as a result V_y causes reactions normal to the tangential and normal to the surface of the element similar to the phenomenon in Figure 7.1. The component of the force δN_y that acts along the axis of propulsion ($\delta N_y \sin(\theta)$) counteracts the retarding effect of all the forces acting tangentially to the surface.

Since dimensions of an element are minuscule and the displacement speed low, the reactions from the medium can be assumed to be directly proportional to the velocity of displacement, and to the viscosity of the medium. The velocity of displacement tangential to the body is $V_y \sin(\theta)$, the tangential drag (δL_y) is $\xi_T V_y \sin(\theta) \delta s$, and the normal force ($\delta N_y$) to the surface is $\xi_N V_y \cos(\theta) \delta s$. ξ_T and ξ_N are the drag coefficients of Equation 7.13. The resultant forward thrust (δF_y) along the axis of propulsion is ($\xi_N - \xi_T$)$V_y \sin\theta \cos\theta \delta s$. This equation shows that the only way the displacement of the element will result in a forward thrust in the axis of propulsion is if $\xi_N > \xi_T$. The sperm is moving not only transversely across the axis of propulsion but also along the latter axis at V_x, which is dependent on how the entire sperm is moving through the medium. V_x also has two components: the tangential displacement $V_x \cos(\theta)$, and the displacement normal to the surface of the element $V_x \sin(\theta)$, which results in the forces acting tangentially and normally to the surface (δL_x and δN_x, respectively). $\delta L_x = \xi_T V_x \cos(\theta) \delta s$, and $\delta N_x = \xi_N V_x \sin(\theta) \delta s$.

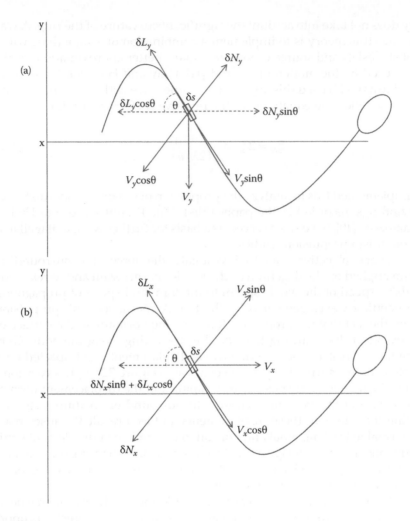

FIGURE 7.2
Representation of Gray and Hancock's swimming sea urchin spermatozoa. (a) Forces impressed on an element δs moving transversely with a velocity of V_y; (b) Forces impressed on the element δs when swimming along the x axis with a velocity of V_x.

The total forces acting on the surface because of the elements' transverse and forward displacements are

$$\delta N = \xi_N \left(V_y \cos(\theta) - V_x \sin(\theta) \right) \delta s, \quad \delta L = \xi_T \left(V_y \sin(\theta) + V_x \cos(\theta) \right) \delta s \quad (7.14)$$

The resultant forward thrust is

$$\delta F = \delta N \sin \theta - \delta L \cos \theta \quad (7.15)$$

since the propulsive components of δN and δL are $\delta N \sin\theta$ and $\delta L \cos\theta$, respectively. So

$$\delta F = \left[(\xi_N - \xi_T)V_y \sin\theta\cos\theta - V_x(\xi_N \sin^2\theta + \xi_T \cos^2\theta) \right]\delta s$$

$$= \left[\left((\xi_N - \xi_T)V_y \tan\theta - V_x(\xi_T + \xi_N \tan^2\theta) \right) \Big/ (1 + \tan^2\theta) \right]\delta s \qquad (7.16)$$

From Equation 7.14, we can see that a forward thrust is possible only if $V_y > V_x \tan\theta$ and Equation 7.16 shows that the forward thrust happens only if $\xi_N > \xi_T$. Now the propulsive speed can be expressed in terms of the wave speed as long as the propulsive speed remains constant during the whole cycle of the element's motion. The above principles can be applied to an undulating filament driving an inert head by setting the transverse velocity of an element (δs) to dy/dt. The tangent of its angle of inclination is dy/dx, so Equation 7.16 becomes

$$dF = \left[\left((\xi_N - \xi_T)\frac{dy}{dt}\frac{dy}{dx} - V_x\left(\xi_T + \xi_N\left(\frac{dy}{dx}\right)^2 \right) \right) \Big/ \left(1 + \left(\frac{dy}{dx}\right)^2 \right) \right]\delta s \qquad (7.17)$$

The total force along the length of the flagellum over one wavelength is

$$F = \int_0^\lambda dF \qquad (7.18)$$

The equation for a filament propelling an inert head is

$$n\int_0^\lambda dF - \xi_H \bar{V}_x = 0 \qquad (7.19)$$

with $\bar{V}_x$ being the average velocity of forward propulsion over a complete cycle of activity, ξ_H is the drag coefficient of the head (for a sphere $\xi_H = 6\pi a\eta$, with a being the radius of the sphere and η being the viscosity of the medium), and n is the number of waves exhibited by the whole tail.

This work has served as a basis for further work in flagellated microorganism propulsion studies. Lighthill [13] has conducted further analysis on this problem and provided more accurate values for ξ_N and ξ_T. Even though there are still limitations to the local drag theory, it has been shown to be useful for calculating and predicting characteristics of flagella-based propulsion of microorganisms. The local drag theory has since been expanded upon by using the slender body theory for a more accurate analysis.

7.3.2 Cilia

Similar to the eukaryotic flagellum, cilia consist of a "9 + 2" structure but are shorter than the flagella. Eukaryotic flagella and cilia are structurally identical and the distinctions between them are often made based on their function and length. Together cilia and flagella are known as a group of organelles called *undulipodia* [14,15]. The motile cilia enable cells to move by moving liquid past the surface of the cell. The cause of cilia movement is similar to that of the eukaryotic flagella; ATP hydrolysis creates forces in the dynein arms that cause the structure to bend. Unlike eukaryotic flagella, cilia perform 3D motion and consist of a power and recovery stroke [16].

7.3.3 Pseudopodia

The next swimming strategy for eukaryotes to be discussed is the pseudo-podia mechanism. Pseudopodia are temporary extensions of eukaryotic cell cytoplasm. They are used by sarcodine protozoans, some flagellated proto-zoans, some cells of animals, and amoeba as a means of locomotion. They extend and contract by the reversible assembly of actin subunits into micro-filaments; projections occur when there is an interaction between myosin and filaments near the cell's end [17]. The projections continue until actin reassembles into a network. In the case of amoeba, the amoeba extends multiple pseudopodia and then drags itself toward the direction of one of them. Four of the best known types of pseudopodia are axopodia, filopodia, lamellipodia, and reticulopodia. Axopodia are characteristically stiff, slen-der, and have internal rods consisting of an array of microtubules. Filopodia are also slender but do not have microtubules (in most protists); instead, they contain actin filaments. The lamellipodia, which are characteristic of *Amoeba*, are broad flat protrusions, and reticulopodia are slender and anastomos-ing pseudopodia that form a cross-connecting net that is stiffened by both microtubules and filaments [18].

7.4 Dynamics Modeling and Analysis of a Swimming Microrobot for Controlled Drug Delivery

This section discusses possible design and control of a swimming microro-bot. We specifically use the work of Li, Tan, and Zhang in their "Dynamics Modeling and Analysis of a Swimming Microrobot for Controlled Drug Delivery" [19,20]. In the paper, a swimming robot composed of a spiral type head and an elastic tail modeled with the resistive force theory previously discussed is proposed. The microrobot is designed for optimal drug deliv-ery, and thus suitable for low Reynolds number swimming. External rotating magnetic fields drive the head of the swimming robot, which allows it to be remotely controlled wirelessly. The head serves as a base for the elastic tail and houses the communication and control units. When a rotating magnetic field is applied, the head rotates in unison with the field, and a driving torque is propagated to the straight elastic tail. Deformation of the tail occurs when a threshold level of the driving torque is reached resulting in the tail transforming into a helix, and generating propulsive thrust. The tail also serves as a drug reservoir in controlled drug delivery operations.

The modeling and analysis of the microrobot were split into analysis before and after the bifurcation of the tail. Bifurcation is the remarkable deformation of rotating filaments [20]. Li et al.'s previous studies had shown that applying increasing driving torque to straight elastic polymers that are clamped at one end and open at the other results in a strongly discontinuous shape deformation at a finite torque (N_C) and a finite rotation frequency (ω_C). In order for significant propulsion to be achieved, the driving torque or rotation frequency has to reach the above thresholds (N_C and ω_C).

7.4.1 Before Bifurcation

Prior to the deformation, the tail is relatively straight and propulsion is negligible, and the tail can be regarded as pure payload [20]. For the entire system to rotate the viscous friction resulting from the head and tail has to be overcome by the driving torque. The resistive torque (T_R) equation is

$$T_R = T_H + T_T \tag{7.20}$$

T_H and T_T are resistive torques from the head and tail, respectively. T_H is then decomposed into

$$T_H = T_P + T_S \tag{7.21}$$

T_P is the resistive torque of the payload, while T_S is the resistive torque from the spiral head. So the moment balance equation of the microrobot is

$$T_D + T_T + T_P + T_S = 0 \tag{7.22}$$

T_D is the driving torque on the head from the external magnetic field, and it is represented as

$$T_D = mH \sin \beta \tag{7.23}$$

where m is the magnetized moment of the magnetized head, H is the magnetic field's amplitude, and β is the angle between them. The payload in the head is modeled as a sphere with radius d. The resistive torque of a sphere according to Stokes' law is

$$T_P = D_R \omega \tag{7.24}$$

with D_R a sphere's rotational drag coefficient,

$$D_R = 8\pi\eta d^3 \tag{7.25}$$

where d is as the radius of the sphere. The tail is modeled as an elongated rod, so the axial rotational resistance torque can be represented as

$$T_T = E_R L_{Tb} \omega \tag{7.26}$$

where L_{Tb} is the length of the tail before bifurcation, ω is the swimming velocity of the microrobot, and E_R is the rotational resistance of an elongated rod

$$E_R = 4\pi\mu\eta d^2 \tag{7.27}$$

a_S is the cross-sectional radius of the elastic tail.

The viscous force per length along the y direction due to the spiral head is assumed to be F_Y. So T_S is represented by

$$T_S = A_H F_Y L_H \tag{7.28}$$

where A_H is the helical amplitude of the head and L_H is the length of the head. So the moment balance equation from Equation 7.22 can be rewritten as

$$T_D + E_R L_{Tb} \omega + D_R \omega + A_H F_Y L_H = 0 \tag{7.29}$$

Similarly, the force balance equation is

$$F_T + F_P + F_S = 0 \tag{7.30}$$

where F_T is the translational resistive force brought by the circular tail, F_P is the translational resistive force brought by the payload, and F_S is the translational resistive force brought by the head itself.

Since the elastic tail is treated as an elongated rod before bifurcation, F_T can be obtained from

$$F_T = E_L V \tag{7.31}$$

where V is the velocity of the microrobot, and E_L is the translational resistance of the elastic tail, and according to [21]

$$E_L = \frac{2\pi\eta L_{Tb}}{\ln\left(\frac{L_{Tb}}{2a_b}\right) + \ln 2 - \frac{1}{2}} \tag{7.32}$$

Next, F_p is defined as

$$F_p = D_L V \tag{7.33}$$

where D_L is the translational drag coefficient of a sphere,

$$D_L = 6\pi\eta d \tag{7.34}$$

The viscous force per length along the x direction due to the spiral head is assumed to be F_X. So F_S is represented by

$$F_S = F_X L_H \tag{7.35}$$

Therefore, the translational force balance equation is

$$E_L V + D_L V + F_X L_H = 0 \tag{7.36}$$

The paper uses resistive force theory (RFT) to approximate F_X and F_Y as long as the radius of the microrobot body is extremely small compared to other relevant lengths.
ξ_L and ξ_N are given by [22]:

$$\xi_T = \frac{2\pi\eta}{\left(\ln\left(\frac{2P_H}{b}\right) - 2.90\right)}, \quad \xi_N = \frac{4\pi\eta}{\left(\ln\left(\frac{2P_H}{b}\right) - 1.90\right)} \tag{7.37}$$

where b is the cross-sectional radius of the wire of the helical head, and P_H is the pitch of the helical head. The local velocities are decomposed into tangential and normal velocities; the relationships between velocities and forces along these directions are

$$F_L = \xi_T (\omega A \sin\theta - V \cos\theta), \quad F_N = \xi_N (\omega A \cos\theta + V \sin\theta) \tag{7.38}$$

where F_L and F_N are the tangential and normal forces, respectively. θ is defined as

$$\tan\theta = 2\pi A_H / P_H \tag{7.39}$$

The forces along the x and y directions are represented by F_L and F_N:

$$F_X = F_L \cos\theta - F_N \sin\theta, \quad F_Y = F_L \sin\theta + F_N \cos\theta \qquad (7.40)$$

The translational velocity of the microrobot can be solved by combining Equations 7.36–7.40 into

$$V = \frac{\omega A_H L_H \sin\theta \cos\theta (\xi_N - \xi_T)}{D_L + E_L - L_H (\xi_N \sin^2\theta + \xi_T \cos^2\theta)} \qquad (7.41)$$

The force along the x direction is

$$F_X = -V(E_L + D_L)/(L_H) = \frac{\omega A_H \sin\theta \cos\theta (\xi_T - \xi_N)(D_L + E_L)}{D_L + E_L - L_H (\xi_N \sin^2\theta + \xi_T \cos^2\theta)} \qquad (7.42)$$

And the angular velocity of the microrobot is

$$\omega = -\frac{T_D}{\dfrac{A_H^2 L_H^2 \sin^2\theta \cos^2\theta (\xi_N - \xi_T)^2}{D_L + E_L - L_H (\xi_N \sin^2\theta + \xi_T \cos^2\theta)}} \qquad (7.43)$$

where $S = E_R L_{Tb} + D_R + A_H^2 L_H (\xi_T \sin^2\theta + \xi_N \cos^2\theta)$.

The system has zero total moment because the driving torque is balanced by the resistive torque from the head and the tail.

7.4.2 After Bifurcation

The elastic tail is assumed to be a circular cylinder with two geometric quantities, radius a and contour length L_{Tb}, and two material quantities: the bending modulus B and twist modulus T. The twist behavior was ignored because it does not directly relate to the bifurcation. Before bifurcation, the angular velocity ω of the microrobot is proportional to the driving torque. When ω is low, the tail remains straight while twisting along its centerline. When ω reaches its bifurcation frequency, the straight elastic tail shape turns into a helix. The bifurcation frequency of the tail is

$$\omega_c = \frac{B}{E_R (L_{Tb})^2} \qquad (7.44)$$

When B is larger, the tail is stiffer and more resistant to bending; when E_R and L_{Tb} are larger, the tail experiences greater friction from the fluid and makes the tail easier to bend.

The torque and force balance Equations 7.22 and 7.30 still hold after bifurcation with one exception: the tail is no longer just payload but is a propulsive unit. The balance equations are

$$T_D + A_T F_Y^* L_{Ta} + D_R \omega + A_H F_Y L_H = 0 \tag{7.45}$$

$$F_X^* L_{Ta} + D_L V + F_X L_H = 0 \tag{7.46}$$

A_T is the amplitude of the tail after bifurcation, L_{Ta} is the length of the tail after bifurcation, F_X^* and F_Y^* and are the viscous force per unit length in the x and y directions due to the elastic tail. V and ω are related by

$$V = P\omega \tag{7.47}$$

where

$$P = \frac{M + M^*}{D_L - N - N^*}$$

$$M = A_H L_H \sin\theta \cos\theta (\xi_N - \xi_T)$$

$$N = L_H \left(\xi_N \sin^2\theta + \xi_T \cos^2\theta \right)$$

$$M^* = A_T L_{Ta} \sin\theta^* \cos\theta^* \left(\xi_N^* - \xi_T^* \right)$$

$$N^* = L_{Ta} \xi_N^* \sin^2\theta^*$$

$$\xi_T^* = \frac{2\pi\eta}{\left(\ln\left(\frac{2P_T}{a} \right) - 2.90 \right)}, \quad \xi_N^* = \frac{4\pi\eta}{\left(\ln\left(\frac{2P_T}{a} \right) - 1.90 \right)},$$

P_T is the pitch of the helical tail, A_T is the amplitude, and L_{Ta} is the length. The asterisk denotes components of the tail after bifurcation. F_Y and F_Y^* in Equation 7.45 can be represented as $Q\omega$ and $Q^*\omega$, respectively.

$$Q = A_H \left(\xi_T \sin^2\theta + \xi_N \cos^2\theta \right) + P \sin\theta \cos\theta (\xi_N - \xi_T)$$

$$Q^* = A_T \left(\xi_T^* \sin^2\theta^* + \xi_N^* \cos^2\theta^* \right) + P \sin\theta^* \cos\theta^* \left(\xi_N^* - \xi_T^* \right)$$

Substituting F_Y and F_Y^* into Equation 7.45, the angular velocity can be written as

$$\omega = -\frac{T_D}{A_T L_{Ta} Q^* + A_H L_H Q + D_R} \tag{7.48}$$

Through simulations, the preceding model of a microrobot was shown to be able to move more freely in a constrained area due to the flexible dimensions and was more efficient when compared to a rigid body spiral-type microrobot [20]. Li et al. also asserted that the elastic tail could lead to easier formation of a flagella bundle which in turn would produce greater thrusts than a single one. Also, because the tail is not just pure payload, it helps generate additional thrust and increase the energy efficiency of the microrobot. Li et al. proposed that this design could be used for a wide range of medical applications.

Acknowledgment

The authors are grateful for the support of the Office of Naval Research under award number ONR-N00014-11-1-0622.

References

1. Batchelor, G.K. *An Introduction to Fluid Dynamics.* Cambridge, UK: Cambridge University Press, pp. 211–215.
2. Cohen, N. and J.H. Boyle, Swimming at low Reynolds number: A beginners guide to undulatory locomotion. *Contemporary Physics*, 2010, **51**(2): pp. 103–123.
3. Purcell, E.M. Life at low Reynolds number. *American Journal of Physics*, 1977, **45**(1): pp. 3–11.
4. Jarrell, K.F. and M.J. McBride. The surprisingly diverse ways that prokaryotes move. *Nature Reviews Microbiology*, 2008, **6**(6): pp. 466–476.
5. Blair, D.F. and S.K. Dutcher. Flagella in prokaryotes and lower eukaryotes. *Current Opinion in Genetics and Development*, 1992, **2**(5): pp. 756–767.
6. Mattick, J.S. Type IV pili and twitching motility. *Annual Review of Microbiology*, 2002, **56**: pp. 289–314.
7. Henrichsen, J. Twitching motility. *Annual Review of Microbiology*, 1983, **37**: pp. 81–93.
8. O'Toole, G.A. and R. Kolter. Flagellar and twitching motility are necessary for Pseudomonas aeruginosa biofilm development. *Molecular Microbiology*, 1998, **30**(2): pp. 295–304.
9. Mcbride, M.J. Bacterial gliding motility: Multiple mechanisms for cell movement over surfaces. *Annual Review of Microbiology*, 2001, **55**: pp. 49–75.
10. Gray, J. and G.J. Hancock. The propulsion of sea-urchin spermatozoa. *Journal of Experimental Biology*, 1955, **32**: pp. 802–814.
11. Lauga, E. and T.R. Powers. The Hydrodynamics of Swimming Microorganisms. Reports on Progress in Physics, 2009, **72**(9) 096601.

12. Hancock, G.J. The self-propulsion of microscopic organisms through liquids *Proceedings of the Royal Society of London Series Mathematical and Physical Sciences,* 1953. **217**(1128): pp. 96–121.
13. Lighthill, J. Flagellar Hydrodynamics: The John von Neumann Lecture, 1975. *SIAM Review,* 1976. **18**(2): pp. 161–230.
14. Gibbons, I.R. Cilia and flagella of eukaryotes. *Journal of Cell Biology,* 1981. **91**(3): pp. S107–S124.
15. Margulis, L. Undulipodia, flagella and cilia. *Biosystems,* 1980. **12**(1–2): pp. 105–108.
16. Walczak, C.E. and D.L. Nelson. Regulation of dynein-driven motility in cilia and flagella. *Cell Motility and the Cytoskeleton,* 1994. **27**(2): pp. 101–107.
17. Sadava, D.E. *Life: The Science of Biology* 2011, Sunderland, MA: Sinauer Associates.
18. Davidovich, N.A. et al. Mechanism of male gamete motility in araphid pennate diatoms from the genus *Tabularia* (Bacillariophyta). *Protist,* 2012, **163**(3): pp. 480–494.
19. Huaming, L., T. Jindong, and Z. Mingjun. Dynamics modeling and analysis of a swimming microrobot for controlled drug delivery. In *ICRA 2006. Proceedings 2006 IEEE International Conference on Robotics and Automation.*
20. Li, H.M., J.D. Tan, and M.J. Zhang. Dynamics modeling and analysis of a swimming microrobot for controlled drug delivery. *IEEE Transactions on Automation Science and Engineering,* 2009, **6**(2): pp. 220–227.
21. Lighthill, J.S. *Mathematical Biofluiddynamics,* 1975, Philadelphia: Society for Industrial and Applied Mathematics.
22. Nakano, M., S. Tsutsumi, and H. Fukunaga. Magnetic properties of Nd-Fe-B thick-film magnets prepared by laser ablation technique. *IEEE Transactions on Magnetics,* 2002, **38**(5): pp. 2913–2915.

8

Modeling and Analysis of the Cellular Mechanics Involved in the Pathophysiology of Disease/Injury

Benjamin E. Reese, Scott C. Lenaghan, and Mingjun Zhang

CONTENTS

8.1 Introduction

Many complex biological processes require continuous multifactorial monitoring, at various time and length scales, in order to elucidate the underlying physiological changes that result over the course of a disease. The technical challenge that this type of monitoring presents is significant to the scientists and clinicians attempting to monitor the pathophysiology of disease. An inability to effectively analyze the events that occur during disease progression further reduces the likelihood of effective therapy and early detection. Additionally, in order to understand complex biological responses, a large quantitative dataset is necessary for analytical analysis, which may provide insight into the underlying mechanisms associated with a behavior or response. Further, the collection of highly quantitative data for biological systems is often costly and time consuming due to the complex equipment and methods associated with their collection. At the single cell level, however, many of the complications associated with in vivo data collection can be avoided. Similarly, in most in vitro studies, cells are not synchronized, leading to variations among the population that contribute to noisy data collection which can be avoided at the single cell level. Ultimately, an understanding of the pathophysiological response at the single cell level may provide insight into disease diagnosis and treatment.

While experimental studies remain the "gold standard" for the development of preventative, diagnostic, and therapeutic strategies, the complex nature of disease relative to the simplicity of experimental studies often leads to the analysis of only a single variable at a time. As a result, it is often beneficial to use mathematical modeling, in addition to biological testing, to gain further insight into these complex processes. The flexibility of models allows them to be used to rapidly generate preliminary data that can lead to the formulation of testable hypotheses in a quick and inexpensive manner. Further, the ability to analyze and manipulate multiple parameters with larger datasets can lead to more comprehensive and robust preliminary investigations, in preparation for further in vitro or in vivo studies. For these reasons, mathematical models are being developed for multiple applications within the fields of biology and medicine to assist in the study of diseases at varying scales ranging from subcellular elements to more complex tissues and organ systems. The following sections illustrate those areas in which the modeling, analysis, and control of cellular mechanics can lead to advancements in the understanding of the pathophysiology of disease.

8.2 Modeling, Analysis, and Control of Cellular Mechanics in Disease/Injury

One of the most difficult aspects of model construction remains in the estimation of model parameters and the identification of the structural and regulatory behavior of biological networks. Generally, this is dependent on the understanding of the system being modeled and the availability of data that can be used to describe the system. Of particular interest to pathologists are changes in the mechanobiology of cells associated with disease. Mechanobiology focuses on how physical forces or changes in cell mechanics contribute to the development and physiology of cells and tissues. Structure–function relationships such as those involved in mechanobiology are known to regulate many biological processes, spanning multiple levels and length scales. As a result, numerous subcellular features, such as those involved in cytoskeletal rearrangement, influence the dynamic behavior of individual cells, which often affects surrounding cells through downstream signaling. This downstream signaling often serves as the origination signal of an injury or disease, and consequently, can be used for early detection of a pathological condition. Subtle modifications in the shape or structure of a cell could represent some of the earliest distinguishable factors indicating the onset of disease. For instance, mechanical forces applied to cells have been shown to regulate the progression of atherosclerosis and influence the transformation from a normal to malignant phenotype in certain cell types [1]. Not only can these external forces directly affect the mechanical response of cells, but they can also trigger the generation or suppression of biochemical and molecular signaling. Both the passive sensing and active modifications exhibited by different cell types due to these forces have an influence on the overall dynamics of cells and tissues. The ability to monitor and quantitatively measure these characteristic changes as a result of patho-logical events could facilitate earlier detection and provide further insight into healthy and diseased states.

Due to the unknown nature and ambiguous interpretation of many bio-logical processes, the ability to address and translate changes from the molecular level to their subsequent physiological influence is a powerful advantage that can be realized with the selection of an appropriate model-ing approach. As a result, models that are capable of integrating components from radically different time and/or length scales are highly sought after. Several examples incorporating multiscale models are accomplished using agent-based modeling. This type of modeling allows for the investigation of the collective behavior of many individual units and the patterns or proper-ties that emerge as a response to certain stimuli [2]. These are typically used when modeling at the tissue level, where changes may have compounding or distinctly different effects at the single cell and systemic level. One recently developed example of using this approach is the Delaunay Object Dynamics (DOD) framework [2]. This framework has been applied to the compartmen-tal homeostasis of B lymphocytes and T lymphocytes in secondary lymphoid tissue. Secondary lymphoid organs are well known for their role in adap-tive immunity, and their cellular organization/migration is considered to be critical for the efficient initiation of the mammalian immune response at these sites. DOD simulations involving contact-dependent forces between cells, such as those that influence topology, have been used to accurately describe these interactions based on Delaunay triangulation [2]. This method allowed for the calculation of several parameters such as the size and posi-tion of cells, which are conditional upon the local environment. Using these values further led to the determination of shape and surface properties, such as adhesion and friction forces, stemming from cellular interactions. The dynamics of this system were described based on Newtonian equations of motion, while the forces involved in these interactions included those pro-vided by the internal cytoskeleton, as well as external forces exerted by the extracellular matrix. Using this type of framework, discrete properties, along with continuous spatial properties, including contact area and the positions of cells, were integrated, allowing the impact of cellular properties on tissue organization to be characterized and explained in the form of a multiscale model. This example illustrates the coupling that exists between cellular mechanics and the organization or function of diverse cell populations in certain tissues. As researchers continue to search for the underlying causes of many diseases and the injuries they can induce, multiscale models will see even further development to increase understanding of the effects of cellular interactions on these complex processes.

Computational modeling can be utilized for the analysis of various cellular processes from both normal and pathological tissue, such as cell migration or adhesion. Epithelial cells, as well as many other cell types, have the ability to form unique shapes and structures in two and three dimensions. Biomechanical properties and the relative forces generated or experienced by individual cells are fundamentally important to understanding how these cells are capable of accomplishing such highly cooperative tasks. Many pathological events occur in response to biomechanical factors and rely on cells' ability to communicate with one another while orchestrating the collective behavior exhibited during these processes. Other examples in which modeling is able to mimic the behavior of individual cells within various tissues was investigated during monolayer formation and tissue morphogenesis [3]. The model used for this study included many different cellular components such as the cytoskeleton, membrane, and cytoplasm, while further accounting for growth, division, motility, apoptosis, and polarization. Simulations using this model were able to show how individual cellular biomechanical properties affected tissue growth and other morphological changes that are critically important to these processes. Experimental studies regarding these processes can sometimes be confined to a two-dimensional substrate or only allow for two dimensions of measurable data, whereas a model such as this could be used to analyze these processes in three dimensions such as that found in vivo. Many times the transition of such a model to three dimensions can be achieved by extending all of the equations to include a third axis. Although this requires a considerable increase in computational cost, it affords the opportunity to study these events in a way that might not be possible using traditional experimental techniques.

Experimentally driven models typically rely on the collection of specific parameter values in order to derive accurate mathematical representations describing a cell's behavior or response. A review of parameter estimation techniques discusses several different approaches for model construction based on this relationship [4]. Traditionally, a forward or bottoms-up modeling approach has been used, where several components are isolated and processed individually. Multiple constituents are then merged together to form an integrative model, such as those used in metabolic pathways, which utilize the kinetic properties of individual enzymes and basic rate laws [5].

Steady-state data are also commonly used in stoichiometric or flux-based approaches in order to estimate parameter values. These types of analyses are usually done by modifying input concentrations for a specified parameter while holding the remaining parameters constant. The effects that these perturbations have on the steady-state values of the system are then studied and recorded for further regression analysis. Due to the reliance of many models on the acquisition of comparable sets of experimental data, the recent advances in biological equipment have had a great influence on the development of cell-based modeling. High-throughput techniques allow for the generation of time series data that can characterize the dynamics of genomic, proteomic, metabolic, and physiological responses for the estimation and identification of important parameters and relationships through the use of a top-down or inverse approach. This type of model construction allows for multiple parameters to be monitored throughout the course of a dynamic biological process, leading to more information on potentially transient or brief deviations occurring at various time points throughout the duration of a study.

These examples help demonstrate how various approaches can be taken in order to apply quantitative measurements to the biological processes being described, such as disease pathology. By integrating a model along with experimental studies, researchers are able to further analyze and relate the growing amount of data that can be collected and used to help characterize and describe these complex biological functions. Modeling these events can also allow for control theory to be incorporated into these studies so that predictive or therapeutic strategies can be applied to enable a more in-depth understanding of how these parameters are related and ultimately provide researchers and clinicians with a clearer picture of how to address or control those diseases.

8.3 Applications in Cancer

Cancer prognosis and treatment have benefited greatly from studies involving gene expression profiles and other clinical or pathological variables. A recent study involving the construction of a prognostic model combining these two approaches showed even further improvements in the accuracy of predictions when compared to either type alone [6]. This model was able to combine two separate methods of prognostication and utilize the statistical analysis of multiple datasets in order to derive a better predictive model that can potentially be used for determining cancer therapy responsiveness and patient outcome. Prognostic variables such as tumor size and genetic signatures were identified and associated with their respective outcomes based on data provided by patients from a combination of multiple clinical trials.

The results from this research show that a model of clinical and genomic variables used in combination with each other had a greater prognostic capability over either predictor individually. This example demonstrates how modeling can combine experimental data spanning different levels of characterization to form a powerful tool for predicting or analyzing complex biological processes such as cancer.

Another example in which modeling has been applied to cellular mechanics involved the quantification of metastatic potential and invasiveness of chondrosarcoma cells [7]. This study developed a thin-layer viscoelastic model for stress relaxation in which the mechanical properties of chondrosarcoma cells of different configurations were quantified using microscopy-based indentation tests with an atomic force microscopy (AFM). The viscoelastic properties of both rounded and spread cell morphologies from several cell lines were acquired and characterized using the derived model to interpret their results. The selection of an appropriate theoretical model for contact geometry is a crucial determinant for the accurate quantification of biophysical properties using this technique. Force-versus-distance curves were taken on each of the different cellular configurations/cell lines and further analyzed using a Hertz-based fit in order to determine Young's modulus. The full viscoelastic response of the cells during stress relaxation tests required a thin-layer correction in order to accurately describe the experimental conditions during the relaxation phase. Darling et al.'s findings were able to demonstrate the effective characterization of different cells using the derived thin-layer viscoelastic model, and further indicated that more aggressive cells exhibited lower moduli. This study was also able to conclude that cell deformability can accurately reflect certain phenotypic characteristics that are related to metastasis, including release from the original tumor site, along with penetration and invasion of the vasculature. Modeling the mechanical properties of these cells led to the confirmation of the hypothesized inverse relationship between metastatic potential and stiffness while providing new insight into therapeutic strategies that might be used to inhibit metastasis by targeting cytoskeletal structures that regulate stiffness and motility.

8.4 Applications in Cardiovascular Disease

As previously mentioned, the role that mechanical forces play in the regulation of atherosclerosis has been recently investigated in terms of the adaptive response of cells in the vasculature under varying flow conditions [8]. Mechanical signaling and mechanotransduction, which describes the molecular mechanisms occurring in response to mechanical cues, are critical determinants of both morphogenesis and function in pathological states such as hypertension and atherosclerosis. Vascular smooth muscle cells

(VSMCs) and endothelial cells (ECs) that line and help form the vasculature are capable of morphological remodeling in order to accommodate shifts in blood pressure and shear stress over relatively long time scales. Chemical cues and other circulating factors are also released upon acute changes in blood flow and the resulting shifts in mechanical forces as an attempt to maintain vascular homeostasis. From this study, it is believed that the ability to adapt to changes helps protect regions of the vasculature from the onset and progression of inflammatory diseases such as atherosclerosis. Other studies have shown that areas in which smooth, laminar flow and vascular homeostasis are maintained lead to a protective effect that decreases EC and VSMC death, promotes cell alignment, and stimulates increased levels of antiinflammatory and antioxidative mechanisms. The extracellular matrix (ECM) has also been shown to undergo changes at atherosclerosis-prone sites of fluctuating mechanical forces by shifting from a normal basement membrane (BM) consisting of collagen and laminin, to proteins characteristic of wounds and inflammation such as fibronectin (FN), fibrinogen (FG), and thrombospondin. Similar studies found that downstream signaling was also affected by these same changes, which then further induced altered gene and protein expression. All of these subsequent changes that occur in response to chronic fluctuations in mechanical cues illustrate the importance of cellular mechanics in relation to the onset and progression of atherosclerotic plaques.

8.5 Advances in Experimental and Imaging Techniques (BioMEMS/NEMS)

Experimental biologists often utilize analytical models by providing measured values for a set of parameters extracted over a time course from a representative study. These are subsequently input into a model to perform simulations and make analytical calculations or predictions based on the data that have been collected. Once validated, models such as these are useful tools because of their ability to identify the most significant factors influencing those changes that are characteristic of a specific disease or injury. From these simulations, parameters having the largest impact can then be targeted and further studied in future experiments to develop more rapid diagnostics and better treatment options.

As made evident by the previous examples, advanced detection and sensing devices can provide a higher level of accuracy to models by acquiring more precise values for measurable parameters. One such field that has emerged due to the increased need to extract similar measurements with improved spatiotemporal resolution from single cell studies is the micro-/nano-fabrication of BioMEMS/NEMS devices. This area has allowed for a higher degree of control and characterization to be applied to, or extracted from, a cell's micro-environment. Cells are most commonly cultured on dishes that are not only stiffer and flatter than most cells' native tissue, but they also lack the appropriate chemical and mechanical signals that cells experience in vivo due to these interactions with micro-environmental cues [9]. A majority of the signaling taking place at the cellular or molecular level during the early stages of disease is initiated before the larger-scale, downstream effects can be detected. In order to develop more efficient drugs, and fully comprehend many of the unknown mechanisms presented by different diseases, it is necessary to address this issue and perform studies using this level of characterization. These devices provide a better understanding of how specific interactions affect cells and cause the changes that take place in response to different stimuli by more accurately representing the local environment at the cellular level.

Some of the primary tools and techniques currently being used for acquiring this type of experimental data are in common use at various levels of characterization [4]. Examples at the genomic level include microarrays, real-time PCR, and RNA-based gene silencing. These allow for the generation of time series data, including expression profiles throughout the course of a study or the system's response to various gene knock-outs. Gel electrophoresis and mass spectrometry (MS) are available for characterization at chosen time points on the proteomic level, while dynamic metabolic studies can also be obtained using nuclear magnetic resonance (NMR), MS, and high performance liquid chromatography (HPLC). These modern techniques provide a clear advantage in terms of the amount of data available for model construction and the ability to extract multiple parameters simultaneously from a single study.

A review of BioMEMS/NEMS illustrates how these devices are improving the depth of understanding for many different biological processes at the cellular level by better representing a cell's native environment [9]. Some of the areas in which improvements have been made due to advances in this field include cell adhesion and traction-force measurements, microfluidics, and micromanipulation. For example, smooth muscle cells lying on an array of microneedles can provide the increased sensitivity and lower detection limit required to measure forces within an appropriate range of nano-newtons [10]. The use of an array also provides higher spatial resolution for the forces generated during the attachment or contraction of cells, lending further knowledge to the force distribution at various locations throughout the cell. In addition to the acquisition of such small forces, AFM cantilevers can be used to either apply or monitor an even broader range of nano-newton forces depending on the selection of an appropriate cantilever and the desired application. Another example utilizes microfluidics in order to simulate the mechanotransduction response of endothelial cells to varying fluid shear stresses and pressures associated with atherosclerosis, allowing for the effects of each parameter to be more closely studied at the cellular level. The development of these and other microfluidic devices have been used to mimic microcirculatory processes, study particle-cell interactions in the targeted delivery of therapeutics or nano drug carriers, and create realistic in vitro models of the microvasculature and its surrounding tissue [11]. Chemical gradients and molecular signaling can also be more precisely controlled and directed through the use of these techniques and devices. The ability to isolate and study many of these events at the single cell level has further advanced modeling capabilities for many of these mechanisms. The use of mathematical models and computational fluid dynamics (CFD) based approaches have also been developed to better interpret many of the complex flow patterns in the microvasculature [11]. BioMEMS/NEMS enable the level of control necessary to re-create and simplify many of the complex networks and reactions experienced during biological processes by isolating and reducing the uncontrollable variables involved in many tissue-level responses.

Because of the added manipulability of BioMEMS/NEMS devices, experiments are able to more closely mimic biological conditions and detect more subtle changes. Microfluidics and micro-/nanoscale fabrication has enabled researchers to better design and re-create environments like those naturally occurring within the body, with the added advantage of being able to direct and manipulate many of the variables that are much more difficult to regulate in vivo. As this field continues to develop, the level of accuracy afforded by cell-based models will also improve and provide scientists with more data and knowledge regarding many complex biological processes.

8.6 An Example: Cardiomyocyte Mechanics

To further illustrate the potential and future directions of this field, an example focused on cardiomyocyte mechanics applied to the study of ischemia and reperfusion injury is proposed in which advanced experimental/imaging techniques and mathematical modeling of single cells can be utilized, followed by a discussion on some of the advantages to this approach.

8.6.1 Experimental Setup/Design

The direct mechanical effects of interactions involving subcellular, cell–cell, and cell–substrate, and those occurring between cells and their local environment have shown their importance in mediating a variety of biological processes. One advantage that has been demonstrated thus far is the fact that these interactions can be described using representative mathematical expressions built on physical principles in order to determine the influence that each parameter has at either the single cell and/or multiple cell levels.

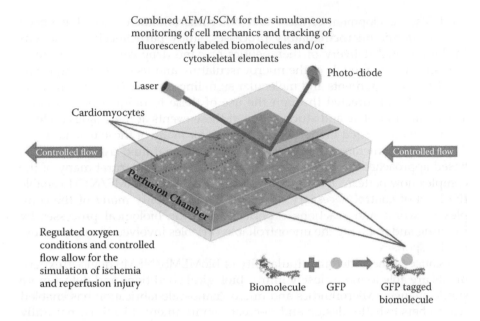

FIGURE 8.1
Experimental setup for the study of ischemia and reperfusion injury utilizing fluorescent monitoring with an LSCM and physical characterization using an AFM for nanoindentation.

In addition to those examples already discussed, the study of cardiomyocyte mechanics is of particular interest due to the unique set of dynamic properties related to their autonomous beating such as stiffness, cytoskeletal arrangement, and a variety of factors including protein or ionic concentration that can be directly related to function and physiology. Advanced imaging used in combination with BioMEMS/NEMS devices or specialized techniques such as nanoindentation can be utilized to acquire enhanced spatiotemporal resolution enabling the characterization and monitoring of events at the single cell or micro-/nanoscale level. An exemplary setup, as shown in Figure 8.1, includes a combined laser scanning confocal microscope (LSCM) for the inter/intracellular fluorescent monitoring of targeted biomolecules or the arrangement of cytoskeletal elements, along with an AFM for nanoindentation that is capable of characterizing the mechanical properties of single cells, such as elasticity or beat-related parameters including force, frequency, and amplitude in real time. The simultaneous acquisition of localized fluorescent intensity and physical properties would then facilitate time-based studies for the examination of pathophysiological events associated with disease or related injuries. Equations derived from indentation can be used to relate these physical parameters to a cell's mechanical properties. Modeling the coordinated efforts of molecular signaling in relation to beating dynamics could provide the information necessary to uncover those mechanisms involved in the transition from normal beating, seen under representative physiological conditions, to arrhythmias or tissue damage seen during ischemic conditions and reperfusion injury. The molecular changes occurring as a result of these states, which can be quantified using techniques such as genomic, proteomic, or metabolomic profiling discussed previously, can also be integrated into a cell-based model through the application of BioMEMS/NEMS devices, further advancing the amount of data and information that can be used to characterize these events. By doing so, not only could a model such as this account for those changes directly related to the physical forces being experienced by the cell, but it could also incorporate the molecular signaling or composition of cells as they constantly undergo changes in response to these fluctuations in a feedback and feed-forward manner. This type of model could then be further expanded to larger, multiscale networks as discussed earlier by involving the interactions of multiple cells with one another and their surroundings. Using this model, one could then look at how modifying single cell or micro-/nanoscale parameter values would affect cell beating and/or its genotype/phenotype from a mechanical and/or biological perspective at a more physiologically relevant tissue level in relation to disease or injury. It is this relationship between the physical and biological features that could ultimately lead to a more in-depth understanding of how a cell functions under various conditions.

8.6.2 Model Development

The development of a cell-based model for the study of ischemia and reperfusion injury should be able to utilize the increased sensitivity and time-dependent data afforded by the proposed experimental setup. A previous model of ischemia and reperfusion was able to describe the relationship between these two conditions based on ionic concentrations of various intra- and extracellular components such as sodium, potassium, and calcium, as well as the inclusion of pH dependence and ATP levels as shown in Figure 8.2 [12]. The total allosteric regulation of the sodium–proton exchange (NHE) was calculated using the following equation,

$$
reg_{NHE} = \left(\frac{\left[H_i^+\right]^{n_{NHE,hi}}}{\left(\left[H_i^+\right]^{n_{NHE,hi}} + K_{NHE}^{n_{NHE,hi}}\right)} \right) * \left(1 - \frac{\left[H_e^+\right]^{n_{NHE,he}}}{\left(\left[H_e^+\right]^{n_{NHE,he}} + K_{NHE}^{n_{NHE,he}}\right)} \right) \quad (8.1)
$$

where $[H_i^+]$ and $[H_e^+]$ represent intracellular and extracellular proton concentration, $n_{NHE,hi}$ and $n_{NHE,he}$ represent the intra- and extracellular Hill coefficients for binding protons, and the dissociation constant was shown as K_{NHE}.

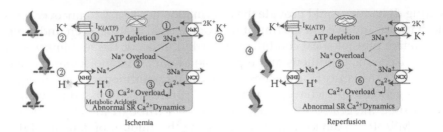

FIGURE 8.2

Models of ischemia and reperfusion as developed by B.N. Roberts and D.J. Christini [12] illustrating the effects of each condition on various ionic exchange pumps in relation to sodium, potassium, and calcium concentrations, pH, and ATP levels. During ischemia, ATP depletion leads to sodium and calcium overload along with increased anaerobic metabolism through various ionic exchange pumps, producing metabolic acidosis, lowered pH levels, and proarrhythmic behavior (1–3). Upon reperfusion, the washout of acidotic fluid reduces extracellular potassium and protons with an increased flux through ionic exchange pumps resulting in a proton gradient exacerbating intracellular sodium and calcium overloads, along with additional proarrhythmic behavior (4–6).

The function for calculating the effects of varying ionic concentrations tied to the amount of water flux in and out of the cell was as follows:

$$V_{H_2O} = 10 * Lp * R * temp * \left(\left(\left[Na_i^+ \right] + \left[K_i^+ \right] + \left[Ca_i^{2+} \right] + \left[Cl_i^- \right] + \frac{\left[X_i^{zi-} \right]}{Vol_{myo}} \right) \right.$$
$$\left. - \left(\left[Na_o^+ \right] + \left[K_o^+ \right] + \left[Ca_o^{2+} \right] + \left[Cl_e^- \right] + \frac{\left[X_e^{zi-} \right]}{Vol_{external}} \right) \right)$$

(8.2)

where Lp is the hydraulic conductivity of the membrane, R is the gas constant, $temp$ is temperature, [Na$^+$], [K$^+$], [Ca^{2+}], [Cl$^-$], and [X^{zi}] represent internal/external concentration levels of sodium, potassium, calcium, chlorine, and impermeable osmolytes, respectively, and $Vol_{myo}/Vol_{external}$ stands for the volume of the myoplasm and external compartment. To simulate the pH changes during ischemia the following equation was used:

$$pH_i = 6.18507 - 0.56698e^{-0.19015t} + 1.5377e^{-0.18462t}$$

(8.3)

describing the changes seen as a function of time, where t is the ischemic time in minutes. Separate equations were also derived to describe the extracellular pH and intracellular anionic changes associated with each time step to include the effects of acidosis as pH decreases. Another marker associated with ischemia and reperfusion includes calcium channel availability with fluctuating ATP levels which was described by

$$f_{Ca,ATP} = \cfrac{1}{1 + \left(\cfrac{k_{ATP}}{[ATP]}\right)^{2.6}} \qquad (8.4)$$

representing ATP concentration with $[ATP]$ as k_{ATP} is the $k_{1/2}$ for the binding of ATP to $I_{Ca(L)}$ channels. Finally, the sarco/endoplasmic reticulum Ca^{2+}-ATPase (SERCA) pump cycling rate was translated to calcium flux using the following equation:

$$V_{SERCA} = 0.00820 \; cyc_{SERCA} \qquad (8.5)$$

where cyc_{SERCA} is the raw cycling rate of the SERCA pump and V_{SERCA} is the output resulting in appropriate Ca^{2+} transients. Through simulations, this model was able to discern one potential cause for the lack of clinical efficacy in NHE inhibition by demonstrating the lack of desired reductions in sodium and calcium overloads during the ischemia–reperfusion event, resulting in the suppression of pH recovery and the inability of the sodium–potassium exchanger to remove sodium from the cell [12]. All of the values and concentrations needed for this model could be accurately characterized and monitored with the use of fluorescence or downstream detection as shown in the specialized platform described previously. Additionally, other signaling molecules such as troponin I, which is known to be released upon myocardial damage, could be integrated into this base model for ischemia and reperfusion in order to monitor the effects that these ionic concentrations, pH changes, ATP levels, and exchange pump rates have in relation to cell damage and the time dependency for the loss of viable tissue in response to the simulated conditions of ischemia and reperfusion.

Along with the molecular and ionic fluxes being represented in association with the dynamics of cardiomyocytes, additional physical monitoring and the acquisition of mechanical properties and dynamics such as cell stiffness, beat force, beat frequency, or beat amplitude, increase the level of characterization available with an advanced platform. The dynamic properties can be acquired by collecting the deflection signal from the AFM. These properties can all be derived from this signal through the following relationship taken from Hooke's law, $F = kx$, where F is the force between the tip and the sample, k is the spring constant of the cantilever, and x is the deflection of the cantilever. The real-time deflection signal using this technique is shown in Figure 8.3, allowing for frequency, force, and amplitude to be derived and constantly monitored throughout the course of a study.

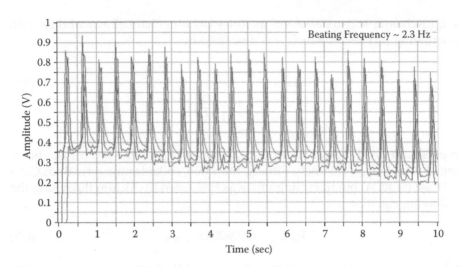

FIGURE 8.3
Deflection signal of a cantilever over a beating cell as shown in Figure 8.1.

In order to measure elasticity, the use of the Hertz theory for calculating the contact area and forces resulting from the tip geometry and cell surface interactions can be applied as previously described [13]. This theory assumes completely elastic behavior and a homogeneous sample. The following example will assume spherical tip geometry and a flat, planar sample surface but can be adapted to various geometries based upon the desired application. Upon indentation and the resulting force-versus-indentation curves, the modulus value can be derived based upon the relationships presented in the following equations:

$$k_s \delta = k_c Z_c \tag{8.6}$$

where k_s and k_c are the stiffness values for the sample and cantilever, Z_c is the deflection of the cantilever and δ is the indentation depth, which is the elastic deformation of the sample calculated by the addition of the deflection signal of the cantilever and the height position of the piezo, Z_p, as shown below.

$$D = Z_p + Z_c + \delta \tag{8.7}$$

where D is the tip-sample separation, assuming an equilibrium once in contact where $D = 0$. Making substitutions, the following equation is obtained:

$$k_c Z_c = -\frac{k_c k_s}{k_c + k_s} Z_p = k_{eff} Z_p \tag{8.8}$$

and $k_{eff} \approx k_s$ when $k_s \ll k_c$. The stiffness of the sample is then related to Young's modulus by

$$k_s = \frac{3}{2} a E_{tot} \quad \text{with} \quad \frac{1}{E_{tot}} = \frac{3}{4}\left(\frac{1-v_s^2}{E_s} + \frac{1-v_t^2}{E_t}\right) \tag{8.9}$$

where v_t, E_t, v_s and E_s are the Poisson's ratio and Young's moduli of the tip and sample, respectively, E_{tot} is the reduced modulus, and a is the tip-sample contact radius. If the tip is much stiffer than the sample and the deformation of the tip is neglected, then this equation can be approximated by

$$k_s = 2a\left(\frac{E_s}{1-v_s^2}\right) \tag{8.10}$$

to describe the elastic deformation of the sample using the Hertz theory. The Hertz model neglects the adhesion between the tip and sample and can be used as an approximation when the adhesion force is much smaller than the maximum load. The contact portion of force-versus-indentation curves are then fit with the Hertz model to obtain the elasticity of the sample. Incorporating this information into the cell-based model, mechanical properties, and other beating parameters can be used to determine the resulting structure–function relationships along with the molecular–biochemical profiles obtained during ischemia and reperfusion during a time study. The ability of the proposed experimental platform to produce this level of characterization and monitoring could enable scientists to uncover the mechanisms responsible for these injuries or other simulated conditions. Quantification from modeling further improves upon the analysis and predictive nature of the system and allows for the study of advanced therapeutic agents or treatment strategies based upon the relationships described within the developed model.

8.6.3 Discussion

The advantages provided by using an advanced experimental platform in conjunction with mathematical modeling for single-cell studies improve upon the current abilities of each approach separately, and further surpass traditional methods in the level of characterization and quantification that can be attained. Single-cell approaches are becoming more relevant as the level of control over experimental conditions improves with the use of BioMEMS/NEMS and other advanced imaging techniques. The ability to monitor changes at the micro- and nanoscale will eventually provide clinicians and researchers with the information needed to uncover the mechanisms involved in complex biological processes. As control over these conditions improves, the computational power associated with the modeling, analysis, and control of these studies will allow scientists to more accurately describe biological functions, further demonstrating the advantages of this combinatorial approach. The current level of characterization described by this proposed setup could help determine the underlying causes of diseases and their associated injuries by combining physical, optical, and chemical techniques with mathematical modeling. While limitations from both modeling and biological perspectives still exist, combining these techniques advances the capabilities currently provided by other methods and opens up the possibility for extending single-cell characterization and monitoring to study complex biological systems and the pathophysiology of disease.

8.7 Conclusions

In conclusion, there are numerous experimental and model-based approaches to investigate the role of cellular mechanics in dictating cellular responses to various conditions associated with disease or other pathologies [14]. Although many of the physiological responses to alterations in these mechanics are known, the difficulty in identifying the underlying mechanisms stems from the inability of most techniques to capture or encompass the changes occurring at both the local (molecular/cellular) and systemic (tissue/organ) levels due to the range of the spatiotemporal domain needed to fully characterize these pathways. Advanced approaches utilizing mathematical modeling and newly developed experimental techniques offer ways to help connect the local and systemic changes associated with these pathologies while affording alternative methods for uncovering the fundamental mechanisms behind these complex biological processes.

Acknowledgment

This research was sponsored by the Office of Naval Research under award number ONR-N00014-11-1-0622. The authors are grateful for the support.

References

1. Hoffman, B.D. and J.C. Crocker, Cell mechanics: Dissecting the physical responses of cells to force, in *Annual Review of Biomedical Engineering*. 2009, Annual Reviews: Palo Alto. pp. 259–288.
2. Beyer, T. and M. Meyer-Hermann, Multiscale modeling of cell mechanics and tissue organization. *Engineering in Medicine and Biology Magazine, IEEE*, 2009. **28**(2): pp. 38–45.
3. Jamali, Y., M. Azimi, and M.R.K. Mofrad, A sub-cellular viscoelastic model for cell population mechanics. *PLoS ONE*, 2010. **5**(8): p. e12097.
4. Chou, I.C. and E.O. Voit, Recent developments in parameter estimation and structure identification of biochemical and genomic systems. *Mathematical Biosciences*, 2009. **219**(2): pp. 57–83.
5. Schallau, K. and B.H. Junker, Simulating plant metabolic pathways with enzyme-kinetic models. *Plant Physiology*, 2010. **152**(4): pp. 1763–1771.
6. Fan, C. et al., Building prognostic models for breast cancer patients using clinical variables and hundreds of gene expression signatures. *BMC Medical Genomics*, 2011. **4**(1): p. 3.
7. Darling, E.M. et al., A thin-layer model for viscoelastic, stress-relaxation testing of cells using atomic force microscopy: Do cell properties reflect metastatic potential? *Biophysical Journal*, 2007. **92**(5): pp. 1784–1791.
8. Hahn, C. and M.A. Schwartz, The role of cellular adaptation to mechanical forces in atherosclerosis. arteriosclerosis. *Thrombosis, and Vascular Biology*, 2008. **28**(12): pp. 2101–2107.
9. Ting, L.H. and N.J. Sniadecki, 3.315—Biological microelectromechanical systems (BioMEMS) devices, in *Comprehensive Biomaterials*, Ducheyne, P., Editor-in-Chief, 2011, Elsevier: Oxford, pp. 257–276.
10. Tan, J.L. et al., Cells lying on a bed of microneedles: An approach to isolate mechanical force. *Proceedings of the National Academy of Sciences USA*, 2003. **100**(4): pp. 1484–1489.
11. Prabhakarpandian, B. et al., Microfluidic devices for modeling cell–cell and particle–cell interactions in the microvasculature. *Microvascular Research*, 2011. **82**(3): pp. 210–220.
12. Roberts, B.N. and D.J. Christini, NHE inhibition does not improve Na^+ or Ca^{2+} overload during reperfusion: Using modeling to illuminate the mechanisms underlying a therapeutic failure. *PLOS Computational Biology*, 2011. **7**(10): p. e1002241.
13. Butt, H.-J., B. Cappella, and M. Kappl, Force measurements with the atomic force microscope: Technique, interpretation and applications. *Surface Science Reports*, 2005. **59**(1–6): pp. 1–152.
14. Loh, O., A. Vaziri, and H.D. Espinosa, The potential of MEMS for advancing experiments and modeling in cell mechanics. *Experimental Mechanics*, 2009. **49**(1): pp. 105–124.

9

Hybrid Control for Micro/Nano Devices and Systems

Xiaobo Li, Xinghua Jia, and Mingjun Zhang

CONTENTS

9.1 Introduction

Regulation mechanisms in biological systems usually employ two basic control structures: feedback and feed-forward [1]. Hybrid feedback and feed-forward control structures are common in biological systems. Biological regulation mechanisms can also be categorized as having either data-driven control or time-driven control. Compared to the prevalence of feedback and time driven controls, theoretical investigations and practical applications of feed-forward and data-driven controls are not new [2–9]. One work was presented in Reference 10, where the inversion-based feed-forward control design was introduced for output tracking of nonlinear systems. It is applicable to multiple-input and multiple-output (MIMO) nonminimum phase systems through calculating the noncausal inverse. For the nonminimum phase systems, a preview-based feed-forward output tracking control was proposed and used for scanning tunneling microscopy control [11]. However, these methods did not take advantage of the feedback measurement. Data-driven control has also been proposed in robotic trajectory planning, regulation of the outputs of biological systems by controlling microenvironement, etc. [8,9,11]. Motivated by the need in gene therapy, a data-driven feed-forward control mechanism that integrates the feed-forward control and data-driven control has been proposed in Reference 12. However, the method proposed is only for single-input–single-output (SISO) minimum-phase systems with continuous measurement. It may not work when the measurement is in a hybrid format, a case that prevails in biological cellular systems and multi-scale engineering systems.

Inspired by regulation mechanisms of biological systems, we propose in this chapter a hybrid time-data-driven control framework for micro-/nanoscale devices or systems with hybrid sensing signals. This is significant for developing a control theory for manipulation of cellular systems and for controlling micro-/nanoscale engineered systems. In this work, we first develop an inverse-based feed-forward control to track the reference trajectory that is generated by either a time-driven or data-driven reference generator (this is due to the nature of biological cellular systems, so that we need the hybrid combination). Depending on the real-time availability of the measurement, the control system is able to switch itself between the data-driven control and the time-driven control.

This study aims to make several original contributions in controlling micro-/nanoscale systems. First, compared to the data-driven feed-forward control proposed previously [12], this new approach can be applied to MIMO biological cellular system control. In addition, compared to the standard time-driven and data-driven approach [12], the proposed control is applicable to measurements with any hybrid format. In principle, each channel can accommodate signals in different formats, including no signal, continuous-time signal, discrete-time signal, and hybrid signal in cases where that signal is either continuous or discrete.

This work is motivated by the need to externally control biological cellular systems for bioengineering applications, where the need is fast emerging from nanomedicine and molecular and cellular engineering. However, there is no control theory available for the purpose. For cellular control, accessible measurement is not always persistent, but often exhibits a hybrid characteristic. In addition, to control biological cellular systems requiring rapid response in nanoscale, a feedback control is not guaranteed to be effective due to relatively slow measurements.

9.2 Problem Formulation

Most micro/nanoscale devices and systems can be modeled as MIMO nonlinear dynamic systems, where partial inputs can be externally manipulated and partial outputs may be measured in real time [1,13]. Here we can use the following generic MIMO nonlinear dynamical model to describe biological systems.

$$\dot{x} = f(x) + g(x)u$$

$$y = h(x)$$

(9.1)

where $x \in R^n$, $u \in R^q$ and $y \in R^q$ are the state, the input, and the output, respectively, and $f(x)$, $g(x)$, and $h(x)$ are smooth vector fields with compatible dimensions. Without loss of generality, we assume that $f(0) = 0$ and $h(0) = 0$. The control objective is to design a control input u, using the available measurement y, such that the output y tracks a reference signal $y_d(t)$. Here, the tracking control problem means that the trajectory of y follows $y_d(t)$ in terms of the data space. This control problem is critically needed in cellular system control [1,13]. We make two assumptions.

Assumption 1. The origin of the linearized system

$$\dot{x} = \left. \frac{df(x)}{dx} \right|_{x=0} x =: Ax$$

is stable.

Assumption 2. The system (Equation 9.1) has a well-defined relative degree $r = (r_1, r_2, ..., r_q)$ at the origin [14].

Remark 1. Assumption 1 is straightforward, since cellular systems are stable in the health state, which was assumed as origin in almost all practices of cellular system modeling. Assumption 2 requires that β be nonsingular. This requirement can be relaxed by adding integrators [40].

Remark 2. Devasia et al.'s paper [10] requires the assumption that the system (9.1) has a minimum phase. However, this assumption is not required in this study. Specifically, if the system (9.1) is a nonminimum phase, the noncausal inverse can be employed to compute the internal dynamics through the integration of a backward differential equation [10].

9.3 Control Framework, Control Design, and Analysis

9.3.1 Control Framework

Figure 9.1 shows the general measurement format. It includes three possible cases: (1) no measurement, (2) continuous measurement, and (3) isolated

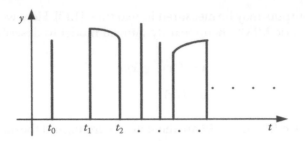

FIGURE 9.1
Illustration of the signal formats of different measurements.

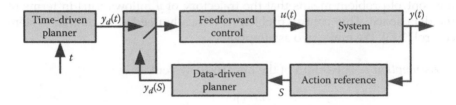

FIGURE 9.2
The proposed hybrid time–data-driven control framework.

measurement. It is common in cellular system control in which persistent measurement is not easy due to the measurement technique. In addition, for a MIMO system with multiple output channels, each channel could have its own measurement format.

The proposed hybrid time–data-driven control framework is shown in Figure 9.2. The inverse-based feed-forward control is to make the output y track the reference trajectory y_d. The reference trajectory is produced by either the time-driven planner or the data-driven planner. The two independent planners are able to be switched, depending on the real-time form of the measurement outputs. When no measurement is available, the time-driven controller will be used. When continuous-time measurement is available, the data-driven controller will be used if certain conditions can be met. Thus, when the measurement arrives in a discrete-time manner, or hybrid form, then the hybrid time–data-driven control is obtained. The data-driven planner is triggered by an event. This control approach covers the classical time-driven and data-driven controls as two special cases.

9.3.2 Feedforward Control Design

The idea is to take the inverse of the original system (9.1) as the feed-forward control [14], based on the exact linearization. Define

$$r = \sum_{i=1}^{q} r_i$$

By Assumption 2, $y_i(\cdot)$ can be differentiated r_i times so that one $u_j(t)$ appears explicitly [10]. Define $\xi_k^i(t) = y_i^{(k-1)}(t)$ for $k = 1, 2, \ldots, r_i$ and $i = 1, 2, \ldots, q$, and denote

$$\xi(t) = [\xi_1^1(t) \quad \xi_2^1(t) \quad \ldots \quad \xi_r^1(t) \quad \xi_1^2(t) \quad \ldots \quad \xi_{r_2}^2(t) \quad \ldots \quad \xi_{r_q}^q(t)]^T$$

$$= [y_1(t) \quad \dot{y}_1(t) \quad \ldots \quad y_1^{(r_1-1)}(t) \quad y_2(t) \quad \ldots \quad y_2^{(r_2-1)}(t) \quad \ldots \quad y_q^{(r_q-1)}(t)]^T$$

Choose η, an $n - r$-dimensional function on R^n, so that $[\xi^T, \eta^T]^T = \varphi(x)$ forms a change of coordinates with $\varphi(0) = 0$. With these new coordinates, the system dynamics (9.1) leads to

$$\dot{\xi}_{r_i}^i(t) = \alpha_i(\xi, \eta) + \beta_i(\xi, \eta)u \tag{9.2}$$

$$\dot{\eta} = s_1^i(\xi, \eta) + s_2^i(\xi, \eta)u, \quad i = 1, 2, \ldots, q \tag{9.3}$$

where Equations 9.2 and 9.3 can be written in a compact form

$$y^{(r)}(t) = \alpha(\xi, \eta) + \beta(\xi, \eta)u$$
$$\dot{\eta} = s_1(\xi, \eta) + s_2(\xi, \eta)u \tag{9.4}$$

where $y(t) = [y_1, \ldots, y_q]^T$, $u(t) = [u_1, \ldots, u_q]^T$, $\alpha(\xi, \eta) = L_f^r h(\varphi^{-1}(\xi, \eta))$ and $\beta(\xi, \eta) = L_g^1 L_f^{r-1} h(\varphi^{-1}(\xi, \eta))$. Since $f(0) = 0$, it follows that $\alpha(0,0) = 0$.

By Assumption 2, β is nonsingular, and then we can define the following feedforward control law:

$$u = \beta^{-1}(\xi_d(t), \eta_d)[y_d^{(r)}(t) - \alpha(\xi_d(t), \eta_d)] \tag{9.5}$$

where $\xi_d(t) := [\xi_{d1}^1(t), \xi_{d2}^1(t) \ldots, \xi_{dr_q}^1(t)]^T = [y_{d1}(t), \dot{y}_{d1}(t), \ldots, y_{dq}^{(r_q-1)}(t)]^T$, and η_d, the states for the feedforward control, can be characterized by the dynamics:

$$\dot{\eta}_d(t) = s(\xi_d, \eta_d, y_d^{(r)}) := s(\eta_d, Y_d)$$

where $s(\eta_d, Y_d) := s_1(\xi_d, \eta_d) + s_2(\xi_d, \eta_d)(\beta^{-1}(\xi_d(t), \eta_d)[y_d^{(r)}(t) - \alpha(\xi_d(t), \eta_d)])$ and $Y_d(\cdot)$ represents ξ_d and $y_d^{(r)}$.

For the general system: $\dot{x} = f(x,u)$ and $y = h(x)$, the above derivation of feedforward control still works, but Equations 9.2, 9.3, and 9.4 will not be in the affine form. Specifically, Equation 9.4 can be replaced by the general form: $y^{(r)}(t) = \alpha(\xi, \eta, u)$ and $\dot{\eta} = s(\xi, \eta, u)$, where u is not explicitly present. Obviously, if the solution for u exists, i.e., $u = \gamma(y^{(r)}, \xi, \eta)$, then the feedforward control can take $u = \gamma(y_d^{(r)}, \xi_d, \eta_d)$. To our knowledge, the general condition for solving this equation is unknown, but the condition for the affine systems is Assumption 2. In addition, the solution exists, if the original system is SISO, which will be demonstrated by the example in Section 9.4.

9.3.3 Design of Time-Driven and Data-Driven Planners

As can be seen from Section 9.3.2, the feed-forward control (9.5) relies on the reference y_d and its derivatives. The reference trajectory can be set or optimized according to practical requirements. For a given smooth trajectory, we focus on the procedure of constructing the corresponding time-driven and data-driven planners to generate the required reference trajectory. This will be further illustrated by an example in Section 9.4.

For the time-driven planner, the reference signal $y_d(t)$ is generated by a reference model:

$$\dot{y}_d = v(y_d(t)), \quad y_d(t_0) = y_{d0} \tag{9.6}$$

where $v(\cdot)$ represents a specific function. With the above equation, we can construct other derivatives $y_d^{(i)}$, all of which form the time-driven signal Y_d to drive the feed-forward control (9.5). For example, consider the following presetting reference trajectory

$$y_d(t) = (y_T - y_0)(1 - e^{-at}) + y_0 \tag{9.7}$$

where y_0 is the initial value for reference, y_T is the final value, and a is a parameter. This corresponds to a linear dynamical system

$$\dot{y}_d = -a(y_d - y_T), \quad y_d(0) = y_0 \tag{9.8}$$

The construction of the data-driven planner has been investigated [14]. To obtain the data-driven planner, an appropriate event must be defined such that all the reference inputs and their derivatives are described by the function of this event. For specific control tasks, we need to choose appropriate events [9,12]. For the above trajectory (Equation 9.7), we define an event $S = y - y_T$, and thus obtain $y_d(S) = y(S) = S + y_T$ and $y_d^{(i)}(S) = -ay_d^{(i-1)} = (-1)^i a^i S$, $i > 1$. All the above expressions are the data-driven signals, which are used to drive the feed-forward control (9.5). Since the data-driven signal relies on the event S that is derived from the measurement, the data-driven planner will lead to a feedback loop, as shown in Figure 9.2. In addition, the time-driven control system can be thought of as evolving under the time reference. A data-driven control system can be thought of as evolving under the data reference.

9.3.4 Planners Switching Control Mechanism

We use a switching control mechanism to switch planners. When no measurement is available in a time interval of (t_i, t_{i+1}), we use the time-driven feed-forward control (9.5). When the measurements arrive at an interval of $[t_i, t_{i+1}]$, the data-driven feedforward control will be

$$u = \beta^{-1}\left(\xi_d(S), \eta_d\right)\left[y_d^{(r)}(S) - \alpha\left(\xi_d(S), \eta_d\right)\right] \quad \text{if} \quad \frac{dS}{dt} > 0 \tag{9.9}$$

At the isolated time instant, the time-driven feed-forward control will be

$$u = \beta^{-1}\left(\xi_d(S), \eta_d\right)\left[y_d^{(r)}(S) - \alpha\left(\xi_d(S), \eta_d\right)\right] \tag{9.10}$$

If at some time instant

$$\frac{dS}{dt} > 0$$

is not satisfied, the time-driven control is still used. Note that in the above forms, $\xi(S)$ and $y_d^{(r)}(S)$ are not meant to replace time t by S, but are driven by the event. In addition, when the control is switched from data-driven control to time-driven control, the time-driven planner is retriggered, i.e., to set a new initial condition. In fact, we let $y_{d0} = y_d(t+) = y(t-)$, where $t+(t-)$ represents the time instant immediately after (before) the jump. It shows one benefit of the proposed control: it is able to adjust the path planner strategy based on the real-time measurement. If there is no measurement, the hybrid control turns out to be a special case of the time-driven feed-forward control (9.5) proposed in Reference 9. If the measurement is available all the time, then the hybrid control becomes the data-driven feed-forward control as proposed in Equation 9.9. The following theorem guarantees the required tracking of the nonlinear system under the hybrid control.

Theorem 1. Consider the nonlinear system (9.1) where the measurement data arrive in a hybrid manner, and the nonlinear system (9.1) is stable with the control (9.5), (9.9), and (9.10), if the dwell time for the switching from no measurement to measurement is large enough. In addition, the output trajectory driven by u_d will locally track the reference trajectory for all the states in the open ball around the origin. ■

Prior to the proof of Theorem 1, we consider a special case where the measurement data arrive at a time series. The following lemma shows that the system (9.1) has the required tracking by the given hybrid time–data-driven controls (9.5) and (9.10), under certain conditions.

Lemma 1. Consider nonlinear system (9.1) with measurements arriving in a discrete-time manner, then if the dwell time (time from one isolated time to the next one) is large enough, it is stabilized by the control (9.5) and (9.10). In addition, the output trajectory driven by u_d will locally track the reference trajectory y_d for all x in the open ball B^n around the origin.

Proof. Assume that the measurement data arrive at a time series, that is, t_0, t_1, t_2, ..., with corresponding outputs $y(t_0)$, $y(t_1)$, $y(t_2)$, ... It means that there is no measurement available at time interval (t_i, t_{i+1}). At this interval, the feedforward control (9.5) corresponds to a nonlinear system

$$\dot{x}_d = f(x_d) + g(x_d)u$$

$$y_d = h(x_d)$$ ■

We define the error signal $e = x - x_d$. We have

$$\dot{e} = f(x) - f(x_d) + \big(g(x) - g(x_d)\big)u$$

$$= \frac{df}{dx}(x_d)e + l_1(x_d)O\big(\|e\|_2^2\big) + l_2(x_d)O\big(\|e\|_2\big)u$$

$$= \frac{df}{dx}(0)e + l_3(x_d)O\big(\|x_d\|_2\big)e + l_1(x_d)O\big(\|e\|_2^2\big)$$

$$+ l_2(x_d)O\big(\|e\|_2\big)u$$

$$= Ae + w(e)$$

where $w(e) = l_3(x_d)O(\|x_d\|_2)e + l_1(x_d)O(\|e\|_2^2) + l_2(x_d)O(\|e\|_2)u$ and $l_1(x_d) = 1/2 \cdot (d^2 f/dx^2)(x_d)$, $l_2(x_d) = 1/2 \cdot (dg/dx)(x_d)$ and $l_3(x_d) = (d^2 f/dx^2)(x_d)$ are smooth and bounded.

Since x_d and u are bounded [10], then $w(e)$ is also bounded. Specifically, for any $\varepsilon > 0$, there exists $\delta(\varepsilon) > 0$ such that $\|e\|_\infty < \delta(\varepsilon)$ and $\|u\|_\infty < \delta(\varepsilon)$, $\|x_d\|_\infty < \delta(\varepsilon)$ implies $\|w(e)\|_2 < \varepsilon\|e\|_2$.

By Assumption 1, we can choose a Lyapunov function $V_t(e,t) = e^T Pe$, where P is the positive definite solution of the Lyapunov equation $PA + A^T P = -Q$ for a positive definite matrix Q. Given that $\varepsilon < \lambda_{min}(Q)/(2\lambda_{max}(P))$, the derivative of $V_t(e,t)$ is

$$\dot{V}_t(e,t) = e^T(PA + A^TP)e + 2ePw(e)$$

$$= -e^TQe + 2ePw(e) \le -[\lambda_{\min}(Q) - 2\varepsilon\lambda_{\max}(P)]\|e\|_2^2 < 0$$

where $\lambda_{\min}$ and $\lambda_{\max}$ stand for the minimum and maximum eigenvalues, respectively. So we have

$$\dot{V}_t(e,t) \le \frac{-[\lambda_{\min}(Q) - 2\varepsilon\lambda_{\max}(P)]}{\lambda_{\max}(P)} V_t(e,t) < 0$$

which implies that

$$V_t(e, t_i + t_\tau) \le e^{\frac{-[\lambda_{\min}(Q) - 2\varepsilon\lambda_{\max}(P)]}{\lambda_{\max}(P)} t_\tau} V_t(e, t_i)$$

if $t_i + t_\tau \in (t_i, t_{i+1}]$.

At each jump instant t_{i+1}, the data-driven control (9.10) is used, which may cause the jump of signals since it may have $y_d(t-) \ne y_d(t+)$. It further may cause the jump of the Lyapunov function $V_t(e,t)$. We assume that there exists a positive scalar κ such that $V_t(e, t_{i+1}+) \le \kappa V_t(e, t_{i+1}-)$. Then, we have

$$V_t(e, t_{i+1}+) \le \kappa V_t(e, t_{i+1}-) \le \kappa e^{\frac{-[\lambda_{\min}(Q) - 2\varepsilon\lambda_{\max}(P)]}{\lambda_{\max}(P)} t_\tau} V_t(e, t_i+).$$

Thus,

$$t_\tau \ge \frac{\lambda_{\max}(P)\ln\kappa}{[\lambda_{\min}(Q) - 2\varepsilon\lambda_{\max}(P)]}$$

implies

$$\kappa e^{\frac{-[\lambda_{\min}(Q) - 2\varepsilon\lambda_{\max}(P)]}{\lambda_{\max}(P)} t_\tau} < 1,$$

which leads to

$$V_t(e, t_{i+1}+) - V_t(e, t_i+) \le \kappa e^{\frac{-[\lambda_{\min}(Q) - 2\varepsilon\lambda_{\max}(P)]}{\lambda_{\max}(P)} t_\tau} V_t(e, t_i+) - V_t(e, t_i+) < 0.$$

Therefore, if the dwell time is large enough, that is,

$$t_\tau \ge \frac{\lambda_{\max}(P)\ln\kappa}{[\lambda_{\min}(Q) - 2\varepsilon\lambda_{\max}(P)]},$$

we have the Lyapunov function decreasing at jump time instants. The error dynamical system is then stable at $e = 0$. The output $y(t) \rightarrow y_d(t)$ as t goes to infinity due to the smoothness of the function of $h(x)$.

Proof of Theorem 1. The hybrid control is to replace the control $u(t)$ with $u(S)$ at the intervals and isolated points, depending on the availability of measurement,

$$\frac{dS}{dt} > 0$$

and sufficiently large dwell time. Consider the same Lyapunov function $V_t(e,t)$ as given in the proof of Lemma 2. We investigate how $V_t(e,t)$ varies with time t. According to the available measurement and the switching strategy, there are five cases for the measurements.

- *Case 1*: The time-driven control (9.5) is used. It happens mainly when there is no measurement. So $V_t(e,t)$ decreases along with time, according to the proof of Lemma 2.
- *Case 2*: Isolated measurement can be obtained and the data-driven control (9.10) is used. In this case, the data-driven control is triggered at the isolated point, i.e., $u(t)$ is replaced by $u(S)$. According to the proof of Lemma 2, if the dwell time is large enough, $V_t(e,t_{i+1}+) < V_t(e,t_{i+1}-)$. Here, dwell time is from the last jump time to the current jump time. The last jump point could be an isolated point or a transition from continuous data to no data (Case 5).
- *Case 3*: Continuous-time measurement is used for the data-driven control (9.9). We use $V_S(e,S)$ to denote the same Lyapunov function under the data reference. Assume the event is S at a time instant t. Then we have $V_t(e,t) = V_S(e,S)$ and

$$\frac{dV_t(e,t)}{dt} = \frac{dV_S(e,S)}{dS} \cdot \frac{dS}{dt}$$

Now we replace $u(t)$ by $u(S)$. It means that for the same error system the time reference is replaced by the data reference. It can be thought of as the special case of Case 1, but under the data reference. So if S increases, then $V_S(e,S)$ decreases, too. Since time is always increasing, S also increases because

$$\frac{dS}{dt} > 0$$

Therefore, $V_s(e,S)$ decreases. As $V_1(e,t) = V_s(e,S)$, we have $V_1(e,t)$ decreases. It means the same Lyapunov function decreases in the time reference.

- *Case 4:* Switching from no data to continuous data. This case can be thought of as the combination of Cases 2 and 3. It may cause the jump of the Lyapunov function, since it may have $y_d(t-) \neq y_d(t+)$. However, according to Case 2, if the dwell time is large enough, then $V_S(e, S_{i+1}+) = V_t(e, t_{i+1}+) < V_t(e, t_{i+1}-)$ at the jump point. In addition, according to Case 3, $V_S(e, S') < V_S(e, S_{i+1}+)$ when $S' > S_{i+1}+$. That is, it decreases immediately after jumping.
- Case 5: Transition from the continuous data to no data. In this case, $u(S)$ is switched to $u(t)$ and the measurement at the transition time is to retrigger the time planner, that is, to reset the initial condition of time planner (9.6) with $y_d(t+) = y(t-)$. At the jump instant, $y_d(t)$ remains the same, i.e., $y(t-) = y_d(t-) = y_d(t+) = y(t+)$. So we have $S(-) = S(+)$ since the event is constructed based on measurements. Thus, $V_t(e, t-) = V_S(e, S-) = V_S(e, S+) = V_t(e, t+)$. There is no change of $V_t(e, t)$ at the transition time instant.

To sum up, for the hybrid measurement, $V_t(e, t)$ decreases along with times for Cases 1 and 3. For Cases 2 and 4, $V_t(e, t)$ may have a jump that may cause it to increase. However, when the dwell time is large enough, we have $V_t(e, t_{i+1}+) < V_t(e, t_{i+1}-)$, which means that it also decreases from one jump point to the next one. For Case 5, there is no change. Therefore, $V_t(e, t) \to 0$, because $V_t(e, t) > 0$. It follows that $e \to 0$ as time goes to infinity. Therefore, the nonlinear system (9.1) under the feed-forward control (9.5) and (9.9) is stable. This further implies that the state $x(t)$ goes to $x_d(t)$. Thus, the output $y(t)$ goes to $y_d(t)$ due to the smoothness of the function $h(x)$.

> *Remark 3.* In our proposed framework, the feed-forward controller is for tracking the reference signal, while the feedback control is for rescheduling the planner for better tracking performance. Since the switching affects the stability of systems, the dwell time should not be too short. Biologically, it makes sense. In this framework, it is for stability and tracking purposes, and is used to realize better tracking via a data-driven planner.

> *Remark 4.* The result is applicable even if hybrid signals are present at different channels. For this case, if measurements for all output channels are available and the given condition is satisfied, the control system is switched to the data-driven feedforward control. Otherwise, the time-driven feed-forward control will be chosen.

9.3.5 Robustness Analysis

Assume the nonlinear system (9.1) is under external disturbances, that is,

$$\dot{x} = f(x) + g(x)u + g_d(x)d$$

$$y = h(x)$$

where $x \in R^n$, $u \in R^q$, $y \in R^q$, and $d \in R^p$ are the state, the input, the output and the disturbance, respectively, and $f(x)$, $g(x)$, $h(x)$ and $g_d(x)$ are smooth vector fields with compatible dimensions. We assume $\|g_d(x)\| < r\|x\|$, where r is a positive constant. Given that $g_d(x)$ and d are unknown, we can design the feed-forward controller for the system (9.1), which is implemented for the above system. Due to the involvement of the disturbance, the tracking performance may deteriorate. However, the following theorem states that under small disturbance, the tracking still can be guaranteed.

Theorem 2. There exist two positive scalars δ_1 and δ_2 such that if $\|d\| < \delta_1$ and $\|y_d\| < \delta_2$, then the output trajectory driven by feedforward control u locally exponentially tracks the reference trajectory y_d for all x in the open ball B^n around the origin.

    Proof. We prove only the time-driven control case. The proof for other cases is similar to that in preceding sections. To define the error signal $e = x - x_d$, we have

$$\dot{e} = \dot{x} - \dot{x}_d$$

$$= f(x) - f(x_d) + \big(g(x) - g(x_d)\big)u + g_d(x)d$$

$$= \frac{df}{dx}(x_d)e + l_1(x_d)O\big(\|e\|_2^2\big) + l_2(x_d)O\big(\|e\|_2\big)u + g_d(x)d$$

$$= \frac{df}{dx}(0)e + l_3(x_d)O\big(\|x_d\|_2\big)e + l_1(x_d)O\big(\|e\|_2^2\big)$$

$$\quad + l_2(x_d)O\big(\|e\|_2\big)u + g_d(x)d$$

$$= Ae + w(e)$$

where $w(e) = l_3(x_d)O(\|x_d\|_2)e + l_1(x_d)O(\|e\|_2^2) + l_2(x_d)O(\|e\|_2)u + g_d(x)d$ and $l_1(x_d) = 1/2\ (d^2f/dx^2)(x_d)$, $l_2(x_d) = 1/2\ (dg/dx)(x_d)$ and $l_3(x_d) = (d^2f/dx^2)(x_d)$ are smooth and bounded.   ■

With Lemma 1 and $\|g_d(x)\| < r\|x\|$, x_d and u are bounded, making $w(e)$ bounded as well. For any $\varepsilon > 0$, there exists $\delta(\varepsilon) > 0$ such that $\|e\|_\infty < \delta(\varepsilon)$ and $\|u\|_\infty < \delta(\varepsilon)$, $\|x_d\|_\infty < \delta(\varepsilon)$ and $\|d\| < \delta_1$ implies $\|w(e)\|_2 < \varepsilon\|e\|_2$.

    By Assumption 1, we can choose a Lyapunov function $V_t(e, t) = e^T P e$, where P is the positive definite solution of the Lyapunov equation

$$PA + A^T P = -Q$$

for a positive definite matrix Q. Given that $\varepsilon < \lambda_{\min}(Q)/(2\lambda_{\max}(P))$, the derivative of $V(e)$ is

$$\dot{V}_t(e,t) = e^T(PA + A^T P)e + 2ePw(e)$$

$$= -e^T Qe + 2ePw(e)$$

$$\leq -\left[\lambda_{\min}(Q) - 2\varepsilon\lambda_{\max}(P)\right]\|e\|_2^2 < 0$$

where $\lambda_{\min}$ and $\lambda_{\max}$ stand for the minimum and maximum eigenvalues, respectively. The error dynamics is thus exponentially stable at $e = 0$. That further implies that the state $x(t)$ goes to $x_d(t)$, and thus the output $y(t)$ goes to $y_d(t)$ due to the smoothness of the function $h(x)$.

9.4 Example

Regulating the glucose utilization of the GAL network in yeast, *Saccharomyces cerevisiae*, externally is a well-accepted test bed for cellular control [10]. The nonlinear model representing the glucose utilization in the GAL network is [15]

$$[\dot{glu}] = T_2([glu],[glu]_e) - \frac{\mu_{glu}[glu]x_{glu}}{k_{glu} + [glu]} - \delta_d[glu] \tag{9.11}$$

$$\dot{x}_{glu} = \sigma_{glu}m_{glu} - \delta_{glu}x_{glu} \tag{9.12}$$

$$\dot{m}_{glu} = \frac{\alpha_{glu} + \varepsilon_{glu}\left([glu]/c_{glu}\right)^b}{1 + \left([glu]/c_{glu}\right)^b} - \gamma_{glu}m_{glu} \tag{9.13}$$

where $[glu]$, m_{glu}, x_{glu}, $[glu]_c$, and

$$T_2 = k_{tr2}x_{glu}\frac{[glu]_e - [glu]}{k_{mtr2} + [glu]_e + [glu] + a_{tr2}[glu]_e[glu]/k_{mtr2}}$$

are the internal glucose, mRNA, and protein, external glucose concentrations, and the transportation function, respectively. The other relevant variables and parameters as well as their (initial) values can be found in [16]. The external glucose concentration $[glu]_e$, an externally manipulated variable, is taken as the control input for external regulation. We decide to regulate the mRNA concentration m_{glu} to a given level (i.e., from 4000 to 8000), where the real-time measurement is practical thanks to the green fluorescent protein (GFP) technique. Then the output equation is $y = m_{glu}$.

By linearization we can show that the above nonlinear model is exponentially stable at a given equilibrium (i.e., the equilibrium point corresponding to $m_{glu}^* = 8000$), which implies that Assumption 1 is satisfied. It is obvious that (9.13) represents $\dot{y}$. Then we can obtain $\ddot{y}$. Define

$$T := \frac{\varepsilon_{glu} b \left([glu]/c_{glu}\right)^{b-1} \left(1 + \left([glu]/c_{glu}\right)^b\right) - b\left(\alpha_{glu} + \varepsilon_{glu}\left([glu]/c_{glu}\right)^b\right)\left([glu]/c_{glu}\right)^{b-1}}{c_{glu}\left(1 + \left([glu]/c_{glu}\right)^b\right)^2}$$

$$N := \frac{\gamma_{glu}\dot{m}_{glu}}{T} + \frac{\mu_{glu}[glu]x_{glu}}{k_{glu} + [glu]} + \delta_d[glu]$$

Then we obtain

$$[glu]_e = \frac{k_{mtr2} + [glu]N + k_{tr2}x_{glu}[glu] + [glu]\ddot{y}/T}{k_{tr2}x_{glu} - N - a_{tr2}[glu]/k_{mtr2} - \ddot{y}/T}$$

The above derivation also demonstrates that the glucose utilization system has a well-defined relative degree, which implies that Assumption 2 is satisfied. According to our discussion in Section 9.3.2, the feedforward control can be constructed as

$$[glu]_e = \frac{k_{mtr2} + [glu]N + k_{tr2}x_{glu}[glu] + [glu]\ddot{y}_d/T}{k_{tr2}x_{glu} - N - a_{tr2}[glu]/k_{mtr2} - \ddot{y}_d/T} \tag{9.14}$$

It is obvious that the internal dynamics is (9.12), which is stable as $\sigma_{glu} > 0$. Thus, the glucose utilization system is a minimum phase. Since the internal glucose concentration $[glu]$ in the controller (9.14) is unknown, we express it in the following form, based on (9.13) (i.e., to replace y and $\dot{y}$ by y_d and $\dot{y}_d$, respectively):

$$[glu] = c_{glu}\left(\frac{\alpha_{glu} - \dot{y}_d - \gamma_{glu}y_d}{\dot{y}_d + \gamma_{glu}y_d - \varepsilon_{glu}}\right)^{1/b}$$

So the feedforward controller (9.14) can be expressed as a function of y_d, $\dot{y}_d$, $\ddot{y}_d$ and x_{glu}.

We take Equations 9.7 as the reference trajectory, in which a represents the regulation speed. Here, $y_T = 4000$. To set a feasible trajectory, we chose $a = 0.01$. We also chose $y_0 = 2000$, which is different from the real initial value for mRNA concentration 4000, because the real initial mRNA concentration is often unknown.

In the first case, we consider the time-driven control without feedback. The reference trajectory and the obtained output are shown in Figure 9.3a. It can be seen that the input to feedforward control $m_{glu,d}$ is the same as the reference trajectory. It can also be seen that the actual mRNA m_{glu} approaches the reference trajectory gradually, and ultimately reaches the setting point 8000. However, since the initial state m_{glu} is unknown from the perspective of the feedforward controller, perfect tracking is not guaranteed.

In the third case, we consider the data-driven control with full feedback. The tracking performance for m_{glu} from 0 to 300 min is shown in Figure 9.3b. Since we use the data-driven planner all the time, the input to feedforward control $m_{glu,d}$ is the same as the output m_{glu}, which means that it can realize perfect tracking.

We assume that the measurement has an isolated point at time instant 30. It also arrives in the continuous-time format from 37.5 to 50, and from 150 to 300. While implementing the proposed hybrid control, we can obtain Figure 9.4a, showing the available measurements at each time instant. Figure 9.4b shows the tracking performance from 0 to 300 min. It can be seen that when the measurement is available, it can realize perfect tracking. For the time instants without the measurement, since the time-driven control is used, it also tracks the trajectory.

9.5 Conclusions

This chapter has proposed a hybrid time–data-driven control framework for controlling micro/nanoscale devices and systems. This is based on the need for developing a control theory for cellular systems to address the challenge in nanomedicine, and cellular and molecular engineering. To limit the drawbacks of the conventional data-driven feedforward control, the proposed control framework is applicable to systems with various types of measurement. Compared to the data-driven feedforward control in the literature, the proposed approach is applicable to MIMO systems. The above makes this work unique for developing a control theory for the emerging needs of nanobiotechnology and systems biology. Simulation results have demonstrated the effectiveness of the proposed control.

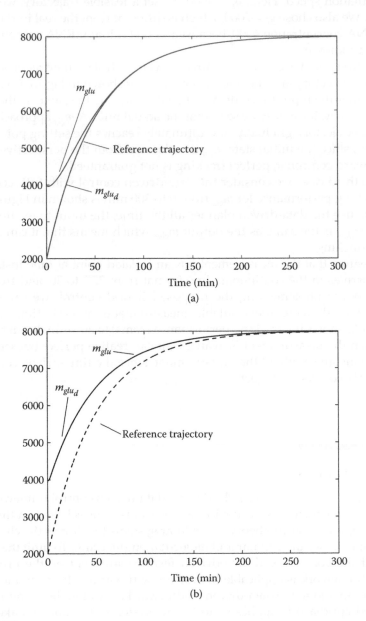

FIGURE 9.3
Tracking performances for time-driven control (a) and data-driven control (b), respectively.

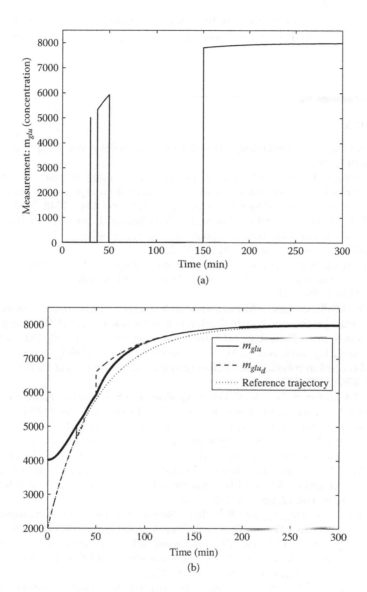

FIGURE 9.4
(a) Available measurement; (b) tracking performance.

Acknowledgment

This research was sponsored by the Office of Naval Research under award number ONR-N00014-11-1-0622. The authors are grateful for the support.

References

1. H. Kitano, "Systems biology: A brief overview," *Science,* vol. 295, pp. 1662–1664, March 1, 2002.
2. M. Treuer, T. Weissbach, and V. Hagenmeyer, "Flatness-based feedforward in a two-degree-of-freedom control of a pumped storage power plant," *IEEE Transactions on Control Systems Technology,* vol. 19, pp. 1540–1548, 2011.
3. L. Brus, T. Wigren, and D. Zambrano, "Feedforward model predictive control of a non-linear solar collector plant with varying delays," *Control Theory and Applications, IET,* vol. 4, pp. 1421–1435, 2010.
4. Z. Hou and S. Jin, "A novel data-driven control approach for a class of discrete-time nonlinear systems," *IEEE Transactions on Control Systems Technology,* vol. 19, pp. 1549–1558, 2011.
5. F. Previdi, T. Schauer, S. M. Savaresi, and K. J. Hunt, "Data-driven control design for neuroprotheses: A virtual reference feedback tuning (VRFT) approach," *Control Systems Technology, IEEE Transactions on,* vol. 12, pp. 176–182, 2004.
6. M. Song, T. J. Tarn, and N. Xi, "Integration of task scheduling, action planning, and control in robotic manufacturing systems," *Proceedings of the IEEE,* vol. 88, pp. 1097–1107, 2000.
7. G. M. Clayton, S. Tien, K. K. Leang, Q. Zou, and S. Devasia, "A review of feedforward control approaches in nanopositioning for high-speed SPM," *Journal of Dynamic Systems, Measurement, and Control,* vol. 131, pp. 061101–19, 2009.
8. N. Xi and T. J. Tarn, "Stability analysis of non-time referenced Internet-based telerobotic systems," *Robotics and Autonomous Systems,* vol. 32, pp. 173–178, 2000.
9. N. Xi, T.-J. Tarn, and A. K. Bejczy, "Intelligent planning and control for multirobot coordination: An event-based approach," *IEEE Transactions on Robotics and Automation,* vol. 12, pp. 439–452, 1996.
10. S. Devasia, C. Degang, and B. Paden, "Nonlinear inversion-based output tracking," *IEEE Transactions on Automatic Control,* vol. 41, pp. 930–942, 1996.
11. Z. Qingze and S. Devasia, "Preview-based optimal inversion for output tracking: Application to scanning tunneling microscopy," *IEEE Transactions on Control Systems Technology,* vol. 12, pp. 375–386, 2004.
12. R. Yang, T.-J. Tarn, and M. Zhang, "Data-driven feedforward control for electroporation mediated gene delivery in gene therapy," *IEEE Transactions on Control Systems Technology,* vol. 18, pp. 935–943, 2010.
13. S. Bewick, R. Yang, and M. Zhang, "Complex mathematical models of biology at the nanoscale," *Wiley Interdisciplinary Reviews: Nanomedicine and Nanobiotechnology,* vol. 1, pp. 650–659, 2009.
14. A. Isidori, *Nonlinear Control Systems,* Third Ed., Springer, London, 1995.

15. R. Yang, S. C. Lenaghan, J. P. Wikswo, and M. Zhang, "External control of the GAL network in *S. cerevisiae*: A view from control theory," *PLoS ONE*, vol. 6, p. e19353, 2011.
16. M. R. Bennett, W. L. Pang, N. A. Ostroff, B. L. Baumgartner, S. Nayak, L. S. Tsimring, and J. Hasty, "Metabolic gene regulation in a dynamically changing environment," *Nature*, vol. 454, pp. 1119–1122, 2008.

Index

Printed and bound by CPI Group (UK) Ltd, Croydon, CR0 4YY

18/10/2024

01776257-0020